GREAT TREES OF BRITAIN AND IRELAND

GREAT TREES OF BRITAIN AND IRELAND

60 of the best ancient avenues, forests and trees to visit

Tony Hall

Kew Publishing
Royal Botanic Gardens, Kew

Royal Botanic Gardens Kew

First published in 2022 by
Royal Botanic Gardens, Kew,
Richmond, Surrey,
TW9 3AB, UK
www.kew.org

ISBN 978 1 84246 746 6

Distributed on behalf of the Royal Botanic Gardens, Kew in North America by the University of Chicago Press, 1427 East 60th St, Chicago, IL 60637, USA.

British Library Cataloguing in Publication Data
A catalogue record for this book is available from the British Library

DESIGN AND PAGE LAYOUT: Kevin Knight
PROJECT MANAGER: Lydia White
PRODUCTION MANAGER: Jo Pillai
COPY-EDITING: Catherine Bradley
PROOFREADING: Michelle Payne

PICTURE CREDITS:
Emma Crawforth: 72, 79, 171; John Delaney: 168; Everyday Artistry Photography / Alamy Stock Photo: 111; Brian Jannsen / Alamy Stock Photo: 83; JaysonTatumGoat (licensed under the Creative Commons Attribution-Share Alike 4.0 International license): 105; Tony Kirkham: 8; Luis Leamus / Alamy Stock Photo: 123; Matthew Scott: 90; Adam Whitbourn: 61.

FRONT COVER: Ardkinglas Silver Fir.

ENDPAPERS: Misty morning in Richmond Park with fallen oak, still very much alive, known as a phoenix tree.

OPPOSITE CONTENTS: The Big Belly Oak, Savernake Forest.

BACK COVER: The Signing Oak in Windsor Great Park.

For information or to purchase all Kew titles please visit shop.kew.org/kewbooksonline or email publishing@kew.org

Kew's mission is to be the global resource in plant and fungal knowledge and the world's leading botanic garden.

Kew receives approximately one third of its funding from Government through the Department for Environment, Food and Rural Affairs (Defra). All other funding needed to support Kew's vital work comes from members, foundations, donors and commercial activities, including book sales.

Printed and bound in Italy by Printer Trento srl.

FSC www.fsc.org MIX Paper from responsible sources FSC® C015829

CONTENTS

ACKNOWLEDGEMENTS

I would like to thank the following for their help and support.

My partner Emma, who (as ever) has trapsed the length and breadth of much of Britain and Ireland for my quest to visit all the trees in this book – during a particularly difficult time as the Coronavirus pandemic made travelling all but impossible at times during 2020-21. My thanks to her also for reading my first drafts.

Laura Martinez-Suz, Research Leader in Mycology at Kew, for checking text and images.

Matthew Scott, a senior ranger at the National Trust, for taking the time to photograph the lime at Florence Court; Adam Whitbourn, head gardener from Blarney Castle, for his image of the western red cedar; and John Delaney, head gardener from Woodstock Gardens and Arboretum for his images of Noble Fir Walk. I was unable to visit these gardens during the pandemic but look forward to visiting in the future.

I would also like to acknowledge the National Trust who, on our behalf, look after and preserve many of the properties and land that feature in this book and protect many of the country's ancient trees, woodlands and forests, which might have disappeared long ago without their protection. Likewise, both English Heritage and Historic Royal Palaces, whose landscapes are home to many wonderful old trees.

Finally, my thanks goes to the team at Kew Publishing, particularly Lydia White, for their help and support, and allowing me to continually change bits!

FOREWORD

Like Tony, my passion for trees started when I was at school. Not being an academic, I was given the responsibility of being blackboard monitor. This meant that every Friday, before the end of the week's classes, my duty was to rid the blackboard duster of chalk dust by running to the end of the playground and banging it against the trunk of a mature horse chestnut, so that it was clean and ready for Monday's lessons. For many years, I remembered this tree as a mighty veteran and a source of seeds for our games of conkers in autumn, only to be disappointed when I returned to rekindle my spirits with this old friend many years later. Despite it being a lot smaller than I remembered, it is still an important tree to me as it was one of the first to kindle my interest in trees and set me on my early career in the world of forestry and arboriculture.

I have been extremely fortunate throughout my working life with trees to be able to travel around the globe in search of new specimens to plant in the Arboretum at Kew. This has given me the opportunity to see many of the species that we grow in the various arboreta in the UK and Kew growing at home in their native habitat, which has greatly enriched my love and understanding of trees. But seeing one growing in the landscape in Britain and Ireland can be just as rewarding and so much easier.

There is nothing more rewarding for any tree or nature lover than meeting an old veteran tree growing well and showing off its personal character – and the more times that you meet these amazing trees, the more you want to learn interesting facts and information about them. When was it planted? Who planted it? What has it witnessed in its time growing? They are also beautiful in their own right.

One of my favourites is the Douglas fir planted at Scone Palace in Scotland in 1827, which has an unusual squat shape with low branching and deeply fissured, dark bark (see page 142). It's not a champion in terms of its size – and there are others in Scotland that are far taller – but it is one of the first Douglas firs grown in Britain, from seed collected by David Douglas in the Pacific Northwest. When I visit this tree, it takes me back to my trips to Vancouver Island in British Columbia and reminds me of my meetings with these arboreal giants in their natural habitat.

A visit to a woodland is good for the soul and mental well-being; in Japan this practice is known as Shinrin-Yoku ('forest bathing'). When in Scotland I always try and pay a visit to the Cairngorms National Park to see some of the old 'Granny' Scots pines. These ancient woodlands are some of my favourites in Britain, with native trees growing together in a very harsh environment.

There are many iconic trees and woodlands with charming character growing in Britain and Ireland, all with interesting and captivating stories behind them. Tony has selected and visited some of his favourites for this beautifully illustrated book, and in my spare time I will now be visiting some of these trees that I haven't yet met and finding out more about why Tony chose them.

Tony Kirkham MBE, VMH

A charismatic 'granny pine' (*Pinus sylvestris*) growing in the Cairngorms National Park.

INTRODUCTION

I am not exactly sure where my initial fascination and love of trees came from. However, I do remember from a very young age seeing fallen acorns under a huge oak tree, and being told that this tree had produced all of these acorns. I was impressed to hear that each acorn, like the one I had picked up and was holding in my small hand, could grow into an oak tree and live for hundreds of years.

Of course, there are trees that from even smaller seeds can attain even greater age, and reach almost unimaginable heights. The giant redwoods (*Sequoiadendron giganteum*) of Northern California, for example, grow to nearly 120 m (400 feet) tall.

I also remember in my early twenties reading a book called *The Man Who Planted Trees* by Jean Giono. It tells the story of a middle-aged shepherd who over decades collects and plants acorns and other tree seeds in a desolate valley in France, eventually turning it into a wonderful green forest. For many years I believed this to be a true story and was in absolute awe of this man for his achievement. I even started to think that maybe I could do something similar. It turns out that encouraging people to plant trees was the author's intention all along!

As an apprentice tree surgeon in the mid-seventies, my first job after leaving school, I saw how Dutch elm disease was sweeping across the country. The impact was terrible: elms were dying in their millions, devastating our native populations and changing the landscape forever, as the elm was one of the most distinctive trees of Britain's countryside.

Autumnal ancient beech trees, some of the magnificent Burnham Beeches.

Some of these trees were huge. I remember one in Ealing Broadway, on the outskirts of London and close to where I lived and worked at the time, whose hollow trunk could easily hold half a dozen people.

Many of my school holidays were spent enjoying adventures in woods and forests. To this day many of my free days and holidays are spent exploring the natural world of both fauna and flora in ancient woodlands, and seeking out remarkable trees, both at home and abroad.

Some are now old friends that I have visited many times over the years and grown to love. Yet first-time visits to new trees are always special. Many are iconic and well documented, with great stories around them – some fact and others fiction, but all there to be enjoyed. Research and studies have proved that walking among trees in woods and forests is good for our health and well-being; they are calming and restful, even lowering blood pressure as you walk among them.

While gathering information for this book, I have learned much more about the trees in our landscape then I already knew, broadening my knowledge and deepening my appreciation of them. There is a great saying that 'every day is a learning day', and I have certainly spent many enjoyable learning days among trees. I love just to stand back and take time to really look at a tree, reflecting on the fact that it began as a tiny seed and now its many branches produce leaves that help create the oxygen we need to breathe. A tree provides food and shelter to countless insects and birds, and is highly important for biodiversity. Below ground there are just as many, if not more, beneficial things going on.

In the UK we have somewhere between 30 and 40 native tree species, depending on how we define 'native' and even a tree! Native trees are generally accepted to be those resident here roughly around 11,000 years ago – that is, after the last Ice Age, and before Britain disconnected from mainland Europe.

Two species of oak, the pedunculate or English oak, *Quercus robur*, and sessile oak, *Quercus petraea*, fall into the native category, as does the yew, *Taxus baccata*; they are our longest-lived native trees. All three can exceed 1,000 years of age. We have more ancient oaks in the UK than anywhere else in Europe and the same can be said for the yew. Other species can also reach this age, but only through continual coppicing.

The seasons bring dramatic changes to the appearance of a tree, transformations that they will go through many times in their lives. Beech (*Fagus sylvatica*) is a wonderful example of this, wearing a coat of many colours through the year. Reddish buds produce fresh, shiny, lime-green leaves in spring, darkening into summer, but it is in autumn that the true glory of these trees appears, their leaves turning fiery tones of golden-yellow, orange and copper before falling in winter to reveal an amazing framework. It is no wonder that the beech is known as the queen of British trees.

For most of my life I have worked with trees, up trees and around trees, planting and helping to preserve them and hopefully raising awareness of them through writing. Trees are such an important part of all our lives; we need them for our own survival, and so must protect

The beautiful form of a natural yew in the Kingley Vale forest.

what we already have. Our ancient trees, woods and forests are truly irreplaceable. Making them more accessible will, I believe, help to protect them. Organisations such as the Woodland Trust, the National Trust and others do a great job in the protection of much of our countryside, but it is also something we can all help with.

Planting and protecting trees for the future is our future too...

Each tree canopy shown here, in Savernake Forest, is giving its neighbour space. This natural occurrence is known as 'crown shyness'.

A rare ancient elm, the survivor of the famous 'Preston Twins', one of which was sadly lost to Dutch elm disease. Mature elms were a common site in the British landscape before the disease took hold in the 1970s.

ANCIENT FORESTS AND WOODLANDS

The forests and woodlands have been an important part of our island and other lands across the world for millennia. Prehistoric people hunted in the forests and woods and used timber for tools, weapons, fuel for heating, lighting, cooking and constructing shelters. Hunting during this period was a necessity of life, and the dense cover created by trees made hunting in forests and woods much easier than on open land. At the same time, open areas of heath and grassland within a forest for deer to graze also provided ideal areas for ambush.

During the Iron Age (C.880 BCE-CE 43), wooded areas such as Hatfield Forest were managed for their timber; animals within them had become semi-domesticated. For example, the domestic pig originates from the Eurasian wild boar (*Sus scrofa*), thought to have first occurred around 9,000 years ago. Life in wooded areas suited boars and swine well; as omnivores, they were able to root around for a wide variety of food including plants and fungi, insects and worms, fallen fruits and nuts. They could also graze smaller fruits and seeds produced on low scrub.

At this time small farmsteads were developing in the open areas in and around the forests. Different types of farming increased, and more and more land was cultivated with the introduction of the iron-tipped ploughshare.

Ancient woodland is defined as woodland that has existed since the seventeenth century in England and Wales, and from the mid-eighteenth century in Scotland. By definition

The tangled mix of live and dead branches of a forest yew in Kingley Vale.

forests are larger than woodlands, although both describe areas covered in trees. The Caledonian Forest in Scotland once covered great swathes of land (possibly as much as 1.5 million hectares (3,706 acres)) in the Scottish Highlands. It is dominated by our only native pine, *Pinus sylvestris*, accompanied by birch, juniper, oak and rowan, and was once the home of beaver, Eurasian lynx, wild boar and grey wolves. The remnants of the Caledonian Pinewoods are still a sanctuary for some of our rarest flora and fauna. Several are very rare or found nowhere else in Britain, for example birds like capercaillie, crested tits and Scottish crossbills and mammals such as the Scottish wildcat, pine martins and red squirrels.

Our oldest forests, especially those in England, are most often associated with hunting, particularly in medieval times. After the Norman conquest in 1066, William the Conqueror introduced Forest Law. This converted most of the land managed under Common Law into Royal Forest, protected for the king as official hunting ground. The 'venison and vert' (deer and trees) now belonged to the king.

Many of the large, mainly deciduous forests that survive today, such as Savernake Forest in Wiltshire and Hatfield Forest in Hertfordshire (one of the oldest hunting forests in Europe), are remnants of land specifically kept for the purpose of royal hunting. Here wild boar, deer and wolves were hunted. Thanks to this early royal protection the trees survived the axe and the land became home to many of the ancient oaks in Britain's forests and woods today. The ancient forests of Scotland mainly survive in rocky, hilly areas that were difficult to farm and colonise.

Veteran maiden oaks, which have never been pollarded, naturally reduce their crown size as they grow into old age.

One of the things that makes ancient woodland so special is that is has developed (in some cases) over thousands of years and is now an increasingly rare habitat. In the last century the UK has lost 50 per cent of its ancient woodland, which now covers less than 2.5 per cent of the country. Over their long existence these irreplaceable forests and woodlands have developed special communities of flora and fauna that can be found nowhere else. Some insects, such as the extremely rare click beetles (*Lacon quercus*), have adapted along with these ancient forests and woods; they can only survive in and on the dead wood of ancient oaks. These woods are also an extremely rich habitat and host to many threatened species of woodland plants, fungi, insects and invertebrates, as well as the microorganisms that rely on them.

Ancient woodland has key indicator plants to look out for. Many are herbaceous, taking advantage of extra light early in the year before the trees come into leaf, creating a cover of sometimes quite dense shade. Examples are wildflowers, including primroses and wood anemones, or swathes of wild garlic, also known as ramsons. And who cannot be transfixed in April and May by the wonderful spectacle of woodland carpeted in our native bluebells (*Hyacinthoides non-scripta*)? This truly iconic wildflower has had many names associated with it over the centuries, such as fairy flower (many folklore tales involve fairies), witches thimbles and English harebells. The UK is the species stronghold and home to almost half of the world's population.

Other woodland plants include ferns, for example the shield ferns *Polystichum aculeatum* and *Polystichum setiferum*, as well as woodland orchids, such as the violet helleborine (with its unusual violet leaves and stems) and the pale, golden flowers of the bird's-nest orchids that thrive among the leaf litter.

Native bluebells (*Hyacinthoides non-scripta*) are often an indicator of ancient woodland. The UK has approximately 50 per cent of the world's bluebell population.

The wild service tree, *Sorbus torminalis*, is a rarity, but may sometimes still be found in ancient woodland. Another tree now rare in woodlands, *Tilia cordata*, the small-leaved lime, is also considered an indicator of ancient woodland. Its sweetly scented flowers form a magnet for bees and its leaves are favoured by one of our commoner large caterpillars, those of the green and yellow-striped lime hawk moth.

Some of these indicator plants will differ, depending on the soil type and what part of the country they are in. Yet generally ancient

woodland habitat offers an undisturbed environment that has developed over centuries. The older the habitat, the more time flora has had to establish itself along with it, increasing the site's botanical importance.

At first glance it is generally the larger, older trees that stand out – but they are only a part of our diminishing ancient woodlands and forests. The action starts below ground with fungal threads and microscopic organisms, then carries on to the smallest invertebrates in the leaf litter, small and large mammals and, of course, the woodland flora. Much of this is rare, and only survives because the woodlands do. I for one hope this is a habitat that will continue for many centuries to come.

Ancient beech pollards in Epping Forest, originally royal hunting grounds.

THE WOOD WIDE WEB

As we wander through forests and woods, a whole new world lies beneath our feet. This world of fungal threads (hyphae), although unfamiliar to most people, is both vast and vital to trees' survival. Under every footstep that we take, there may be as many as 100 km (62 miles) of these threads. They join individual trees of the same and different species together to work as a single organism, helping and benefiting each other. This underground community spreads much further than the trees' own root systems. It interlinks the trees that grow together by attaching and penetrating their root tips to form mycorrhizas (fungus-roots); from these the hyphae spread into the soil, making extensions to the root systems into a superhighway connection.

This complex system has been named 'The Wood Wide Web', or nature's social network. It creates fungal connections between the trees that grow in a wood and is made up of mycorrhizal fungi – beneficial fungi that form biological pathways to connect individual trees. Mycorrhizal fungi have a symbiotic relationship with trees. The latter trade sugars from the photosynthesis with the fungi for soil and water and nutrients, which mycorrhizal fungi can absorb much more efficiently than the tree roots alone.

Many of these nutrients are locked up in the soil and are thus difficult for trees to access on their own. Mycorrhizal fungi are in fact better soil scavengers than the fine roots, so their assistance is valuable. This mutualistic

Beneath the carpet of leaves on the forest floor, another world exists – a world that joins trees throughout the woods via a fungal network enabling them to communicate with each other.

relationship of cooperation and sharing helps both trees and fungi to obtain the elements required for healthy growth.

Trees share up to 25 per cent of the energy that they capture via their canopies from sunlight, through the process of photosynthesis; they convert carbon dioxide and water into sugars, which the fungi use for their growth. The mycorrhiza fungi thus depend on trees for the carbon, transported down through their trunks into their roots, to be taken up by the mycorrhizas. The fungal filaments then extend from the mycorrhizas into the soil, increasing the surface of absorption of the roots up to tenfold and contributing to carbon sequestration in soil.

It is not only the sharing of carbon and nutrients that is beneficial, but also of information. By using the connection provided by mycorrhiza, trees can transmit defence signals. For example, if one tree is being attacked by a predator that is eating its leaves, such as caterpillars or aphids, this information can be passed from tree to tree as chemical stress signals though the underground fungal network. The connection thus warns other trees linked to the network of the attack, enabling them to raise their defences and produce bitter compounds to make their leaves less palatable.

Another example involves trees such as walnuts, which contain a chemical in their leaves called juglone. This is known to suppress the growth of some plants around the tree's area of leaf fall. Juglone is produced by the roots and spread by the network, eliminating local competition.

Scientists also believe that the older and taller trees, sometimes known as 'mother trees', use this network to share their sugars and nutrients with younger, smaller and shaded trees. This helps those in need to survive in more difficult positions.

'The Wood Wide Web' is most active and thrives in areas of unworked woodlands and ancient woods – places in which the soil has been left undisturbed for centuries. In such settings an extensive underground network can link trees in a forest together, young and old, helping all to flourish as one giant super-organism.

Ectomycorrhizas formed on the roots of beech trees. This symbiotic relationship between the roots and the mycorrhiza helps with the plants uptake of mineral, nutrients and water in exchange for carbohydrates.

COPPICE AND POLLARDS

Coppicing and pollarding are both traditional, ancient methods of woodland management, and both take advantage of the natural process by which trees grow. It is through these practices of continually cutting back tree species for a variety of uses that some of our oldest trees have survived, as coppiced stools and pollards – a condition that extends their lives far beyond their normal span.

Coppicing and pollarding mainly differ in the height of cut that is made. Coppicing involves cutting back almost to ground level. In pollarding the tree's growth is cut back to the main trunk but at varying heights, usually around head height.

They are both ways of producing renewal wood for building, fuel, fodder and tools, as well as for many other purposes. Coppiced woodland is mentioned in the Domesday Book, completed in 1086, in which it is called silva minuta, but these skills can be traced back much earlier, over thousands of years. In the 1940s and 1950s prehistoric wooden trackways dating to the Late Bronze/Early Iron Age were uncovered during peat digging in the Somerset Levels. The main body of the trackway consisted of roundwood (trunks and limbs) and brushwood, mostly birch, but also alder; the main structure was made of oak logs and ash poles. All are ideal for coppicing.

Ancient woods that have been used in coppicing have boundary earthworks, usually a bank and ditch. These structures, known as woodbanks, mark boundaries and make a barrier for grazing animals. The bank is built up from the soil excavated from the ditch, so

Looking up among the towering limbs of a lapsed beech coppice.

the ditch is the actual boundary. Woodbanks are not always straight; irregular forms are the oldest, dating from medieval times or earlier. Veteran trees often grow on top of and along these old woodbanks – but evidence of any coppice that would once have existed in the woods within the banks is long gone. It is always interesting to come across woodbanks, however, and to wonder what an industrious place it once might have been.

Trees managed in this way must be able to produce vigorous new epicormic shoots. Not all of them can do this; some will die back if their main (apical) stem is cut back too hard. Many of our native broadleaved species of tree do respond to coppicing, but at differing strengths and speeds. Those with the strongest regrowth include ash, hazel, small-leaved lime, sweet chestnut and willow. Beech and oak also coppice well, but are more suited to pollarding. Conifers, with the exception of yew, will die if cut back to a stump.

Coppicing is the practice of continually cutting a tree back to the base during the dormant season, producing what is known as stool. The stools are entirely self-renewing; they burst into life and produce new shoots from dormant buds. These buds are stimulated by a change in hormone levels when the top growth is removed, during the lengthening days and warmth of late spring; this transforms trees that appeared to have very little life into flourishing wonders of nature. Over the course of just one season the growth can be phenomenal, particularly with hazel and birch. This practice, which can be carried out with basic hand tools, also creates a diversity and mosaic of woodland habitats – brilliant for supporting a huge range of our native flora and fauna.

Coppice is cut on a rotation system that divides areas of woodland into what are known as coupes. These are grouped in blocks, with each coupe being cut sequentially. For example, hazel is usually cut at between 7–10 years, depending on its use. If your woodland was divided into eight coupes each could then be cut on a yearly rotation, giving a crop to harvest, and the cycle restarted. There are two main rotations for coppicing different species: short rotation (for harvesting in less than 10 years) and long rotation (for coppice grown for more than 20 years).

Within many of the old coppice woodlands, standards (uncoppiced trees) were encouraged to grow. In fact, during Henry VIII's reign it was a legal requirement that at least 12 standards were to be grown in every acre of coppice. Standards were grown among the coppice over a longer rotation, providing timber for construction or furniture. They grew tall and straight, with very few lower side branches, suppressed by the 'underwood' or lower growing coppice.

Coppicing has enjoyed a resurgence in recent years due to its sustainable management of woodland and woodland materials. It creates a haven for a rich diversity of wildlife, including rarer mammals such as the dormouse, and carpets of wildflowers in the spring. In a renewable, sustainable world, coppicing has a lot to offer. I for one hope that this traditional practice will continue to grow in popularity.

Pollarding is the process of beheading a young maiden tree. This is done by cutting

An ancient oak pollard.

through the main stem or trunk and then cutting back to the regrowth at this point. The work is undertaken at a regular determined period during the dormant season.

The main reason for pollarding as opposed to coppicing is that the new shoots it produces are above the browsing height of animals – mainly deer, but also domestic livestock such as cattle and horses. Pollards are normally produced above head height at around 2–2.5 m (6.5–8 feet).

Pollarding is thus a practice used in areas that are regularly grazed. An example of this is wood pastures – areas of woodland in which animals, both wild and domesticated, could roam and graze. Between the pollarded trees, grass and other herbaceous plants fed livestock, which in turn would also eat and discourage the regeneration of woody seedlings. The pollarded trees produced shelter and shade too. During autumn livestock, particularly pigs, would be put out in the wood pasture to feast on acorns and beechmast, a tradition known as pannage. This was once a common practice all over Europe and had been since at least medieval times, but it is no longer carried out as it once was. In the New Forest in Hampshire, however, it still takes place each autumn. The pannage season in the New Forest lasts for 60 days. In this time the pigs, as well as being fattened up, eat nuts which are poisonous to the cattle and ponies that roam the forest.

The poles grown on pollards, like those in coppicing, are cut at different ages, depending

These old beech pollards would once have made up an area of wood pasture, pruned above the height of browsing cattle and deer.

on what the regrowth will be used for. The young foliage of pollards was also used as winter fodder for livestock.

Many of the ancient and veteran pollarded trees we see in forests today have not been managed for wood production for many decades. These trees, known as lapsed pollards, are often visually stunning, but may now be difficult to manage. Branch attachments become stressed, for instance, as their weight is held above often decaying, hollow trunks. Lapsed pollards are home to many invertebrates, however, some of which can only survive on the decaying wood they produce. Nesting birds and sheltering bats also make use of their hollows.

These wispy-topped pollarded willows are a familiar sight along streams and rivers.

OVERLEAF
The ancient 'hollow way'
of Mill Lane, in West Sussex.

PROFILES OF TREES, AVENUES AND FORESTS

Tree Avenue Forest

- 1 Ardkinglas Silver Fir
- 2 Balmerino Chestnut
- 3 Baobab Plane
- 4 Barn Elms Plane
- 5 Beauly Elm
- 6 Bedhamton Yews
- 7 Benmore Botanic Garden – Redwood Avenue
- 8 Bicton – Araucaria Avenue
- 9 Bishop's Palace Plane
- 10 Black Wood of Rannoch
- 11 Blarney Castle Western Red Cedar
- 12 Brighstone Dragon
- 13 Buckden Towers Planes
- 14 Burnham Beeches
- 15 Capon Tree
- 16 Chirk Oak
- 17 Clumber Park – Lime Tree Avenue
- 18 Cluny Redwoods
- 19 Cowdray Park Chestnut
- 20 Croft Castle – Chestnut Avenue
- 21 Dark Hedges – Beech Avenue
- 22 Drummond Castle – Beech Avenue
- 23 Epping Forest
- 24 The Florence Court Lime
- 25 Fredville Park Avenue
- 26 Glen Lyon Ash
- 27 Glenmore Forest
- 28 Hethel Old Thorn
- 29 Holker Great Lime
- 30 Kew Black Locust
- 31 King Oak
- 32 Kingley Vale Yews
- 33 Kings Inns London Plane
- 34 Kingston Lacy – Beech Avenue
- 35 Majesty Oak
- 36 Maplebeck Yew
- 37 Meikleour Hedge – Beech Hedge
- 38 Mottisfont Plane
- 39 Muckross Yew
- 40 Plas Newydd Beech
- 41 Posso Sycamores
- 42 Preston Twins
- 43 Queen Elizabeth Oak
- 44 Roslin Chestnuts
- 45 Saltram Sycamore
- 46 Savernake Beech Avenue
- 47 Savernake Forest
- 48 Scone Dougls Fir
- 49 Sherwood Forest
- 50 St John the Baptist Church – Yew Avenue
- 51 Stiperstones Hollies
- 52 Tolpuddle Sycamore
- 53 Tortworth Chestnut
- 54 Warehorne Ash
- 55 Westonbirt Lime
- 56 Windsor Great Park
- 57 Wistman's Wood
- 58 Woodstock Gardens and Arboretum – Noble Fir Avenue
- 59 Wooland Yew
- 60 Wyndham's Oak

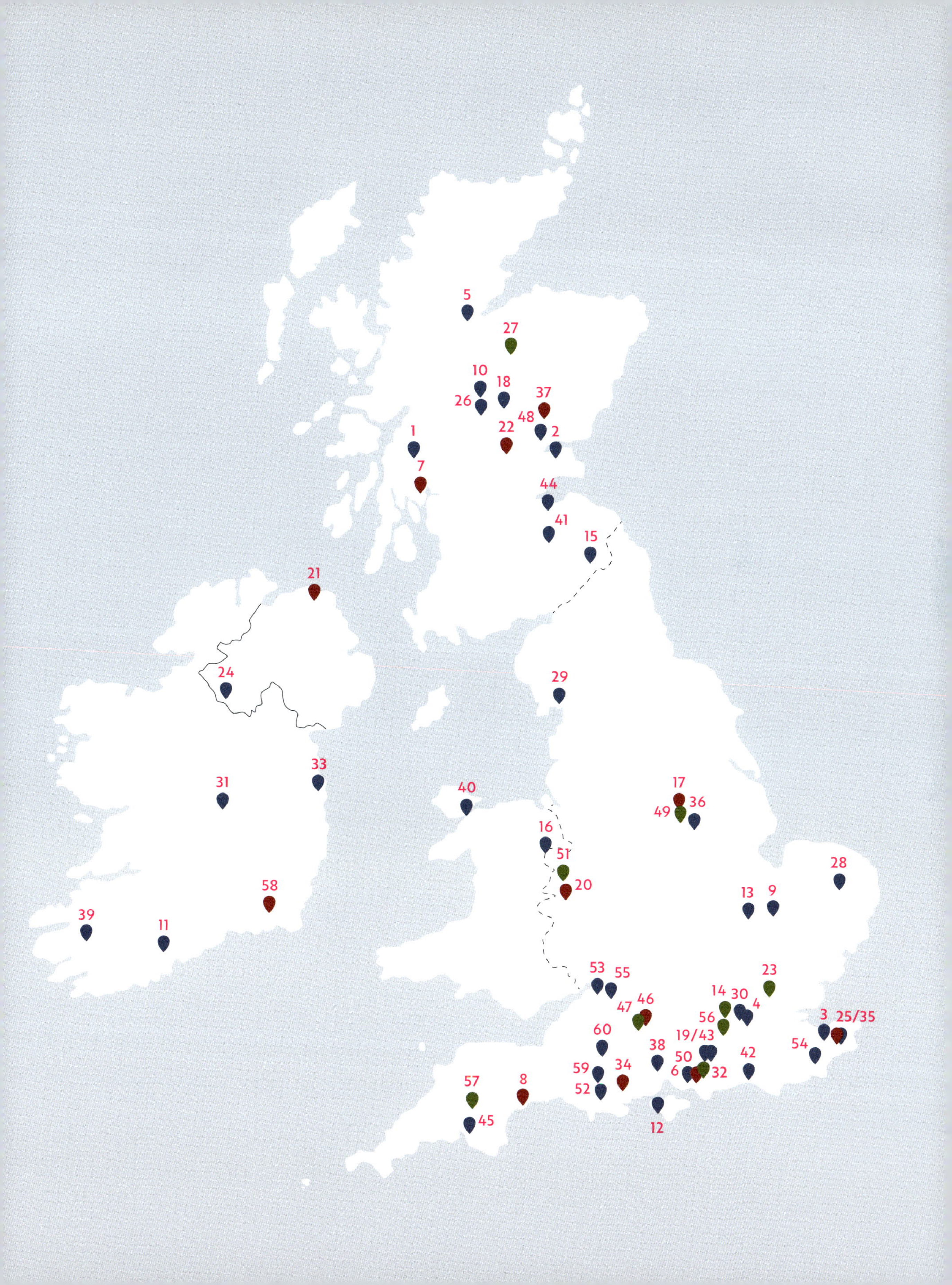

5
27
10
18
26
37
48
22
1
2
7
44
41
15
21
24
29
33
31
40
17
49
36
16
51
28
58
20
13
9
39
11
53
55
23
14
30
47
46
4
56
3
25/35
60
19/43
38
42
54
50
6
32
59
34
57
8
52
45
12

1. ARDKINGLAS SILVER FIR

Champion of Europe

This European silver fir (*Abies alba*) is a tree of gigantic proportions. In the early twentieth century it was dubbed 'the mightiest conifer in Europe' and, like a true champion, it still retains that title today.

Growing in the lower Ardkinglas Woodland Garden, the fir's towering limbs extending from an impressively huge trunk are spellbinding – it will definitely stop you in your tracks. Believed to have been planted in 1750, the first recorded measurement of the tree's girth was in 1905 when it was 6.5 m (21 feet) round and 35 m (144 feet) tall. Its most recent measurements have increased to an enormous girth of over 10 metres, 10.5 m (34 feet), while its height over those years has risen to 48.5 m (159 feet). Definitely the largest conifer in the UK, it is almost certainly the biggest in Europe too.

The Ardkinglas Silver Fir is an unusual shape, resembling a goblet. Its short, thick, deeply fluted trunk divides into four smaller, but still impressive, individual trunks. Each would make a substantial tree on its own. The reason for the unusual form of the fir is not known, but is most likely the result of damage early in life.

The combination of a sheltered position and plenty of soil moisture from an average yearly rainfall of 254 cm (100 inches) provides ideal growing conditions. This part of the woodland garden also features other impressive and notable conifers. Close by grows what was until fairly recently the tallest tree in Britain, a grand fir (*Abies grandis*) at around 62 m (203 feet). It held the title for more than ten years and is still one of the UK's tallest grand firs. Douglas firs (*Pseudotsuga menziesii*) in various Glens of Scotland have now assumed the title, however, reaching heights of more than 66 m (217 feet). This state of affairs remains fluid, however, so in time Ardkinglas may again have both the mightiest and tallest trees in the UK.

LOCATION
Ardkinglas
Woodland Garden,
Cairndow,
Argyll,
PA26 8BG

OS GRID REF
NN179103

The immense towering limbs and girth of this silver fir are astonishing. A true champion, this fir is probably the largest in Europe.

2. BALMERINO CHESTNUT

Growing old gracefully

Beyond the ruins of Balmerino Abbey is an ancient sweet chestnut (*Castanea sativa*). The tree is locally known as Queen Ermengarde's Chestnut, from a tradition that it was planted by Queen Ermengarde in 1229 to mark the foundation of the abbey. She was the widow of William the Lion, a king of Scotland, and was buried in front of the high altar after her death in 1233. Ermengarde's grave, later desecrated, is now marked by a large wooden cross.

The Balmerino Chestnut is the oldest tree in Fife and one of the oldest sweet, or Spanish, chestnuts in Scotland. The spot where it grows is wonderfully tranquil, with nearby views over the Tay estuary and, just behind, the picturesque remains of the Cistercian monastery (a daughter house of Melrose Abbey, the first Cistercian monastery in Scotland). The abbey lands were gifted to the National Trust for Scotland by the Earl of Dundee in 1936, making them the custodian of the chestnut, a living historic treasure.

The tree has a wonderfully sculptural form, beautifully silvered with lichens. Areas of sun-bleached deadwood with longitudinal splits add character to an already charismatic tree. Its bulbous base leads up to two large limbs, twisting away from each other; both are heavily propped. The larger of the two limbs has a cavity filled with concrete, a common practice in the 1960s; it sought to stop the ingress of water and slow down the tree's decay. The chestnut appears almost lifeless in winter, but its appearance is totally transformed in spring and summer with a healthy, shiny green canopy.

It would be great to think that this old timer, supported on its wooden and metal posts, was planted way back in the thirteenth century, making it nearly 800 years old. However, a plaque by the tree states that in September 1988 the National Trust for Scotland, as managers of the site, took 'core borings which indicate it is between 400 and 435 years old'. This is still very old for a sweet chestnut, confirming that the tree is a remarkable ancient survivor.

LOCATION
Balmerino Abbey,
Balmerino Village,
Newport-on-Tay,
DD6 8SB

OS GRID REF
NO357247

Metal props aid the beautiful twisted form of the Balmerino Chestnut. The burred, lichen-covered trunk of this ancient sweet chestnut is also testament to its age.

3. BAOBAB PLANE

Age before beauty

This striking form of the London plane (*Platanus* x *hispanica*) is surely the closest we will get in our temperate climate to the giant baobab trees of Madagascar. It has nothing like the usual form of London planes seen so commonly in Victorian plantings, particularly as avenues of street trees in towns and cities.

The Baobab Plane is a quite remarkable shape. The enormous, round and knobbly trunk does not taper from ground level, as you might expect from a London plane. Instead it forms a wide, squat body that ends abruptly where long, tentacle-like limbs erupt, giving the tree the appearance of some science fiction creature from a 1950s film.

The tree's portly trunk is rumoured to have swallowed up a circular bench that once surrounded the tree – but there is no real evidence of this.

Although these trees are unusual, in Canterbury they are fairly common; at least six others in the city share this same distorted shape. Another one growing in the grounds of the cathedral has a girth of almost 7.5 m (25 feet), while in the churchyard of St Gregory's Church there are two more, flanking the north and southeast corners of the church building. These three trees are thought to have been planted by William Masters (1796–1874), a Victorian botanist and landscape designer. He would presumably would have been responsible for the planting of all the 'baobab planes' in Canterbury.

The distinctive malformed trunks of the baobab planes may be the result of an unknown viral infection that caused this abnormal growth pattern. The trees also often produce short, twiggy and leafy, almost sessile growth directly from their trunks.

LOCATION
Westgate Gardens,
St Peter's Street,
Canterbury,
CT1 2BQ

OS GRID REF
TR146580

The intriguing Baobab Plane looks more like a creature from another world, or some tentacled monster, than a London plane tree (*Platanus* x *hispanica*).

4. BARN ELMS PLANE

'Barney' of London

This little-known London gem is a tree that towers above all around it. Growing in a small, wooded area close to the towpath of the River Thames is a London plane (*Platanus* x *hispanica*). A circular plaque by the tree says 'The Barn Elms Plane. The Oldest in London. One of the Great Trees of London'. Thought to date from around 1685, this is one of the oldest London planes in the UK.

Locally known as 'Barney', this is possibly the largest tree in Greater London. It is an absolute giant, over 35 m (115 feet) in height and with a girth of approximately 8.5 m (28 feet). Its huge knobbly trunk divides into four equally huge limbs that tower skyward; they can be seen standing well above other trees in the wood. A short way above the main trunk the limbs are held together by two metal rods, forming a cross. Above those are two sets of chains, also working to cable the limbs together and hopefully to stop them breaking away from each other and the tree.

The small woodland that remains around the plane tree was once part of a larger estate, the property of the church and Chapter of St Pauls Cathedral for many centuries. In 1579 Elizabeth I bought the lease, which included the manor house known as Barn Elms, as a gift for Sir Francis Walsingham, the queen's 'Spymaster' and Secretary of State.

The manor house was later owned by the Hoare family and in the mid-eighteenth century Sir Henry Hoare developed a garden there. On the edge of the existing woodland is a stretch of water that he extended to create a more formal lake and surrounding area. It still exists as a fishing lake and landscaped garden, including the part where the plane tree stands.

So it is possible that 'Barney' was indeed planted around this time. Yet it may also have been a planting from a former owner, brought into the landscaped garden as a substantial tree. No longer a part of a formal garden, the small woodland (and London's largest plane tree) is now mainly enjoyed by locals; it also provides a campsite for groups of guides and scouts.

LOCATION
Queen Elizabeth Walk, Barnes, London, SW13 9SA

OS GRID REF
TQ229765

Mostly hidden in a small, wooded area, the Barn Elms Plane is given away by its huge canopy that stands head and shoulders above all around it.

5. BEAULY ELM

A Romantic ruin

The ancient Beauly Priory wych elm (*Ulmus glabra*), also known as the Scots elm, is believed to be more than 800 years old. It is the oldest in the UK and may be the oldest in Europe.

Standing all alone and just about clinging on to life, this ancient elm was once part of an avenue of elms that led from the Priory into the town square. This single tree now stands like a spooky sentinel at the gated entrance to the graveyard.

The Priory itself, founded in around 1230, was home to a Valliscaulian order of monks. Now a substantial ruin, its remains are protected as a scheduled monument. The Priory was clearly already a ruin in the early nineteenth century. Keats vividly described it in the first few lines of the poem 'Some Skulls in Beauly Abbey, Near Inverness', written in 1818:

In silent barren Synod met,
Within those roofless walls where yet
The shafted arch and carved fret
Cling to the ruin,

Covered in burrs and knobbly bark (much of which has started to fall away), the elm looks every bit its age, now many centuries. Branches cut back in an attempt to increase its life and stability have become hollow stubs, some without bark, surrounded by the tree's final attempts to produce new life in twiggy branches to continue its growth. These too are now all but dead.

As with many of elms across the UK, Dutch elm disease has taken its toll on this ancient tree. Today only one small area on the top of its trunk still shows signs of life, with just a few leaves struggling on, so sadly its days are inevitably numbered. The tree will remain though, as long as it is safe, as it is such a significant symbol in the town.

As with many ancient trees, the end of the individual's life does not mean all is lost. The Beauly Elm will continue to give life to other organisms. Mosses and lichens have started to colonise its warty trunk and many invertebrates rely on the elm's dead and decaying wood. Its hollowed branches provide ideal shelters for roosting bats, nesting birds and overwintering insects.

Standing close by the iconic elm, and now the most dominant tree in the graveyard, is a huge sycamore (*Acer pseudoplatanus*) more than 200 years old. It has a silvered, deeply fissured and burred trunk that

LOCATION
Beauly Priory,
Beauly,
Inverness,
IV4 7DY

OS GRID REF
NH527464

Sharing the graveyard of Beauly Priory with the ancient elm is this huge, equally impressive, silver-trunked sycamore, itself over 200 years old.

measures around 5.5 metres (18 ft) in girth and an enormous, well branched canopy.

This poem about the sycamore was written by Primary One, Beauly Primary School:

Our sycamore tree
An ancient tree
A humongous, bumpy, lumpy tree,
A seven-children-with-arms-out-
stretched-wide tree
A wrinkly, rough, gnarly tree,
A tangled, twisted, swaying branches tree,
A twirling, swirling, twisting, helicopter seed tree,
Standing tall amongst the graves.

The gnarled, burr-covered trunk of the Beauly Elm, one of oldest elms in Britain. Now barely alive, it is the last survivor of an avenue that led from Beauly Priory into the town.

6. BEDHAMPTON YEWS

The odd couple

Two ancient yew trees dominate the small churchyard of the beautiful St Thomas's Church, which dates from the twelfth century and still retains some of its medieval fabric. One is a male tree and the other, slightly larger, a female. As well as being different sexes, the trees are also completely different in their forms.

The trunks of both yews are of similar size, with girths of approximately 7 m (23 feet). Both have some hollow areas in their trunks, as is usual for ancient yews, and both have had these cavities filled with bricks and concrete over the years. Once a common practice, intended to keep out water and prevent further decay, this is now known to have a detrimental effect on the trees, reducing their ability to move and flex.

The male yew on the south side of the church has a deeply fluted trunk that is quite uniform in its width; it is slightly wider at its base, but not by much. One of its lower limbs was clearly lost some time ago. Just above the lost limb, at around 2.5 m (8 feet), the trunk divides into the main limbs that make up its canopy.

The female yew, close by and to the east of the church, is the more interesting. It is slightly larger, leaning and growing at a concerning angle of almost 30 degrees. Seemingly defying gravity, it produces most of its growth on the back side, away from the lean, in a conscious effort to maintain its stability. A large, re-purposed wooden telegraph pole serves as a prop against its lean; I am sure its support is helping the tree.

The wide lower trunk of the female yew is almost completely hollow, although from one side this is not very apparent. The hollow has been filled with what looks like a badly built wall, cobbled together with brick in an attempt (well-meant at the time) to strengthen and protect the hollow trunk. The rippled bark almost appears to be flowing over the brick interior, a waterfall frozen in time.

Around the tree's base is a low oval wall, made up of broken paving slabs. This strange feature adds to the quirkiness of this strikingly unusual yew.

LOCATION
St Thomas's Church,
Bidbury Lane,
Bedhampton,
Havant,
PO9 3JG

OS GRID REF
SU702063

Entering the churchyard of St Thomas's Church in Bedhampton, you are greeted by two ancient yews. This larger leaning yew has a deeply fissured trunk and a token wooden prop.

7. BENMORE BOTANIC GARDEN – REDWOOD AVENUE

Avenue of giants

When you pass through the gates of Benmore Botanic Garden in Scotland, you are greeted by the Redwood Avenue, an avenue of giants. Without doubt the best-known avenue of trees in Scotland and possibly in the UK, it is surely the most spectacular entrance to any botanic garden.

The first few of these redwoods, on the garden's entrance side of the small bridge, serve as an appetiser at the beginning of the avenue. You then cross over the small bridge (with sequoia cone finials) and the River Eachaig, which flows through the botanic garden on its way to Loch Eck. As you do so, the dramatic, main avenue of giant redwoods (*Sequoiadendron giganteum*), commonly known in the UK as Wellingtonias after the Duke of Wellington, stands proudly ahead of you.

Twenty pairs of redwoods were planted in 1863 by the American owner of the Benmore estate, James Piers Patrick. One tree was replaced in 1995, but the remaining giants are now nearly 160 years old. With heights of over 50 m (164 feet), with the tallest almost 57 m (187 feet), and girths ranging from over 5 m to almost 7 m (16 to 23 feet), they are among the oldest and tallest redwoods in Britain, or even Europe. They form a truly amazing avenue of towering giants, dwarfing visitors who walk along its 350 m (1,150 feet). The redwoods originally lined the driveway to Benmore House, a nineteenth-century listed manor house. Today it is an activity and adventure centre and a facility for outdoor learning holidays.

There are giant redwoods in America that are much older and taller; they have lived for thousands of years and reached heights of over 90 m (300 feet). It is hoped that these youngsters at Benmore, benefiting from high rainfall and a favourable climate in the west of Scotland, may continue to grow – and in future centuries even compete with their American cousins.

LOCATION
Benmore
Botanic Garden,
Dunoon,
Argyll
PA23 8QU

OS GRID REF
NS142855

The spectacular Redwood Avenue at Benmore Botanic Garden consists of towering giant redwoods in proud parade.

8. BICTON – ARAUCARIA AVENUE

A puzzling avenue

The entrance leading up to what is now Bicton College is home to what is claimed to be the longest avenue of monkey puzzles (*Araucaria araucana*) in Europe. When the avenue was planted in 1844, it led up to the Georgian mansion owned by Lord Rolle. His head gardener, James Barnes, was responsible for the avenue's original planting. The trees that form the avenue are thought to have been grown using seed collected in Chile by William Lobb in 1840. Now endangered in its native habitat, *Araucaria araucana* was declared the national tree of Chile in 1976, when its logging was made illegal.

William Lobb, a Cornish plant collector, was not the first to bring *Araucaria* to the UK. In 1795 the botanist and naval surgeon Archibald Menzies, aboard HMS *Discovery* (Captain Cook's old ship) on a voyage circumnavigating the globe, was invited to attend a banquet hosted by the Governor of Chile. *Araucaria* seeds, traditionally harvested by the Mapuche people, were served as part of the meal. Menzies secretly slipped some into his pocket and grew them on the deck of the ship. These plants were the first araucarias brought to Britain, and two of them were planted at the Royal Botanic Gardens, Kew. Sadly, the last of these original trees died in 1892.

The avenue at Bicton is 500 m (1,640 feet) in length and in March 2021 it was made up of 61 trees. Of these, less than 20 are likely to be survivors from the original planting. I suspect that when the avenue was originally planted the trees would have been in matched pairs. This is no longer the case, as many trees have been replaced over the years. The tallest tree is the 'champion' monkey puzzle in the UK, reaching over 27 m (88 feet) in height and possessing a girth of 4 m (13 feet). The younger trees interspersed with veterans along the avenue are grown from seed collected from the older avenue trees, so continuing their genetic lineage. This task is aided by horticultural students from the college.

Endangered in its native habit, *Araucaria araucana* was declared the national tree of Chile in 1976. Its logging was also made illegal.

In Chile the monkey puzzle tree is known by the common name *pewen*. Its common name in the UK, monkey puzzle, apparently derives from a comment made in the mid-nineteenth century about a tree growing in the gardens at Pencarrow, owned by Sir William Molesworth. He was showing a group of friends around the gardens when one of them, on seeing this tree for the first time, remarked 'it would puzzle a monkey to climb that'. It seems the name monkey puzzle has stuck ever since.

LOCATION
Bicton College,
Budleigh Salterton,
Devon,
EX9 7BY

OS GRID REF
SY069866

Some of the araucarias in this avenue at Bicton date back to the original planting in 1844. They are believed to have been grown from the first seeds to reach Britain.

9. BISHOP'S PALACE PLANE

Gifted plane

The Bishop's Palace originated in the fifteenth century, incorporating an earlier deanery dating from the thirteenth century. Much of this building was added to and refurbished in the seventeenth century after the ravages of the Civil War. This was when the gift of a London plane tree was planted in the newly created garden.

Now almost 350 years old, the London plane (*Platanus* x *hispanica*) in the gardens of the Old Palace is a tree of absolute gigantic proportions. It is believed to have been planted in 1680 by Peter Gunning, then Bishop of Ely. However, an old black and white photograph is annotated with a claim that it was planted in 1674.

Entering the garden, the tree's impressive scale is remarkable; it dominates its space yet fits within it comfortably. As you approach the huge trunk its enormity seems even more impressive. A girth of over 10 m (33 feet), knobbly roots growing above ground, radiating out a further 3 m (10 feet) around the trunk, and a towering height of over 30 m (99 feet) combine to make it one of the largest and oldest London planes in the United Kingdom.

The old black and white photograph, probably dating from the early twentieth century, shows the tree with four huge limbs. Sadly, the plane now has only three, but each of these has an individual girth larger than the trunks of most street trees. The loss of the other limb has left a large cavity on one side of the tree. Despite this, this veteran plane is still growing healthily, its lofty canopy appreciated by the many crows that find it a safe, if rather windy, nesting site.

Two huge oriental plane trees (*Platanus orientalis*) also grow along the boundary wall inside the garden. The pair may date from a similar period to the London plane.

The Bishop's Palace is managed by King's Ely as a co-educational independent day and boarding school. Founded in CE 970, it is one of the oldest schools in the world. The garden of the Old Palace is home to the school's sixth form centre, so the Bishop's Palace plane can only be visited by appointment or on the annual Heritage Open Days.

LOCATION
Old Palace,
Palace Green,
Ely,
Cambridgeshire,
CB7 4EW

OS GRID REF
TL539803

This enormous London plane tree dominates the garden of the Bishop's Palace at Ely. Its equally huge root plate spreads out into the surrounding lawn.

10. BLACK WOOD OF RANNOCH

An ancient lineage

The Black Wood of Rannoch is a remnant of the ancient and vast Caledonian Forest that once covered much of Scotland. The characterful Scots pines (*Pinus sylvestris*) that survive today are descendants of the first pines that reached Scotland after the last Ice Age, around 11,000 years ago. The Romans called Scotland 'Caledonia', meaning wooded heights, alluding to the pine-covered highlands. It is from this that the old Caledonian Forest gets its name.

Today the Black Wood is the largest remaining area of relict Caledonian Forest in Perthshire. Parts of the forest are particularly dense and made excellent hideaways for illicit stills, producing illegal alcohol. The owners are said to have protected their stills by leaving out a saucer of whisky for the fairies.

In the forest's more open areas are found some of the most impressive of the ancient 'granny pines'. Here they have space to spread their canopies, surrounded by junipers and bilberries (*Vaccinium myrtillus*), cowberries (*Vaccinium vitis-idaea*) and swathes of purple heathers towards late summer. Wildflowers abound along track edges and in the more open spaces, but it will always be the stunning old 'grannies' that steal the show, their trunks protected with thick, scaly plates, deeply fissured by age. They have huge, branched crowns with reddish bark on the higher limbs lighting up their dark canopies, redder when moistened by rain and mist. The mist adds an extra eeriness to these ancient relic pines, part of an unbroken line that reaches back millennia.

The pines of the Black Wood are enhanced by beautiful mature specimens of the downy birch (*Betula pubescens*), whose elegant drooping branches are clothed in hanging lichens and mosses. The birches' grey-green foliage forms a striking contrast with the darker green pines; they grow side by side along the slopes of the Allt na Bogair gorge above the rushing waters of Carie Burn on its way to Loch Rannoch.

As you wander through this beautiful woodland you can imagine being watched by some of the rare and wary creatures that still survive – pine martens, red squirrels and the elusive wild cats. All were once much more common than they are today. The pinewood is also home to Scotland's only endemic bird, the Scottish crossbill, while centuries ago it would also have sheltered beavers, brown bears, lynx, wild boar and wolves. These remnant forests of pine have a unique ecosystem of flora and fauna found nowhere else in the British Isles.

LOCATION
Carie,
Perth and Kinross.
Nearest postcode
PH17 2QJ.

OS GRID REF
NN608562

Some of the giants of the Black Wood of Rannoch, part of the ancient Caledonian Forest.

11. BLARNEY CASTLE WESTERN RED CEDAR

The elephant tree

This unusually shaped western red cedar (*Thuja plicata*) grows in the grounds of Blarney Castle and Gardens. The species, native to the Pacific Northwest, has great cultural significance and many traditional uses for the region's Indigenous peoples. British Columbia adopted the western red cedar as its official tree in 1988, in recognition of its value and importance to First Nations. William Lobb, a commercial seed collector, introduced the species to Britain and Ireland in 1853.

The thujas are thought to have arrived at Blarney around the same time as the current mansion was built (west of the remains of the keep of the fifteenth-century castle). The mansion was completed in 1874, the probable planting date of this tree and the other large western red cedars in the grounds. Now almost 150 years old, they are among the oldest red cedars in all of Britain and Ireland.

The four huge limbs of this tree grow out from the very short trunk, approximately 1 m (3 feet) in height. Two of them head skyward, reaching around 33 m (108 feet), while the other two grow out almost horizontally. It also has a much smaller limb that has layered itself to produce a separate tree, although this remains joined to the original trunk. Where trees layer themselves out from the main trunk in this way, they are often referred to as 'walking trees'. The yew forest at Kingley Vale has many fine examples (p 107). As a result, this thuja is affectionally known as the 'Elephant Tree'. One limb is shaped like the trunk of a huge elephant, complete with an eye-like mark towards the base of the 'trunk'.

The gardens also contain other amazing trees, including one of the oldest yews in Ireland. Voted Irish Tree of the Year in 2109, this yew is thought to be around 600 years old. If you visit, take the time to see – and kiss – the 'Blarney Stone', which bestows the gift of eloquence...

LOCATION
Monacnapa,
Blarney,
Co. Cork.
Nearest
postcode
T23 YA34.

OS GRID REF
No reference
found

Often less is more... but in this case, it is the multiple trunks that make this tree so special.

12. BRIGHSTONE DRAGON

'Sleeping dragon'

Sprawling across a small stream close to Brighstone Mill on the Isle of Wight is a sleeping dragon: an oak many centuries old. The twisting and snaking branches of this oak (*Quercus robur*) span a small stream called Buddle Brook. One of the main branches has come to rest on the opposite side of the stream from its almost prostrate trunk, forming an arching bridge.

Buddle Brook once powered the 200-year-old mill in Brighstone, which closed in the 1960s. A grist mill, it was mostly used for grinding animal food, but did produce some coarse flour as well. The old mill, now a private dwelling, is downstream of the fallen oak. Past the mill, the brook runs through Grange Chine, a coastal gorge, before reaching the coast and disappearing through the shingle on the beach.

A local legend claims that this old oak was once a mythical winged dragon. It terrorised the village of Brighstone, devouring the children, until a brave knight recently returned from the Crusades heard of their plight and came to the villagers' rescue. Battling with the dragon, the knight struck a fatal blow. Upon this the dragon turned to wood and put down roots, transforming himself into this wonderful, twisting oak.

This is a great story, and I would love it to be true (apart from the eating children bit!). More realistically, the tree is almost certainly windblown, knocked over by wind during a storm. No records exist to establish when this might have taken place, but growing close to the edge of the brook would have made the oak unstable. Soil around its roots would have been continually washed away as it matured, enabling strong winds to topple it far more easily. The large arching limb that now supports the oak will have prevented the tree's roots from being completely severed from the ground, allowing the oak to stay alive and continue to grow into this spectacular tree, distinguished by its unusual, somewhat unnatural, resting position. Trees like these, which have enough of their roots intact to survive, are often referred to as Phoenix trees.

The Brighstone Dragon is reached by passing through a kissing gate on to a public footpath close to the junction of Wilberforce Road and New Road, Brighstone, on the Isle of Wight. There is a public car park close by.

LOCATION
Greystones,
New Road,
Brighstone,
Newport,
Isle of Wight
PO30 4BP

OS GRID REF
SZ425824

A dragon in the heart of the Isle of Wight, its long, twisting limbs bridging a small stream.

13. BUCKDEN TOWERS PLANES

Hiding giant

These are regarded as the oldest London planes (*Platanus* x *hispanica*) growing in the UK. The two trees that grow in the grounds of Buckden Towers, formerly the Palace of the Bishop of Lincoln, are very different from one another. Planted between 1660 and 1663 by Robert Sanderson, then Bishop of Lincoln, they are thought to have been gifts from Oxford Botanic Garden. Sanderson had studied at Oxford before becoming Bishop of Lincoln in October 1660.

The smaller of the two trees sits in an open grassy area of the 6-hectare (15-acre) estate, close to the back of the main house. It does not have the girth or height of the larger tree, but it does possess a wide, spreading canopy from low down. The more monumental of the two trees is tucked away in a small, wooded area at the north end of what is now one long pond, but was originally a series of four small fishponds. It may be partially concealed in the small, wooded area, but the sheer size of the tree makes it very difficult to hide.

At well over 30 m (99 feet) tall it stands head and shoulders above everything around it, making it visible from quite a distance. Walking into the woods surrounding this distinguished tree, you see it rising from the small mound on which it sits. Its knobbly, burred and fluted trunk, nearly 8 m (26 feet) in girth, divides into four equally sized limbs at a height of around 4 m (13 feet) high. These seem to keep on heading skywards, dividing again and again to form a huge crown.

The actual origins of the London plane are unknown. In the mid-seventeenth century, botanist John Tradescant the Younger grew both of this tree's parents, the oriental plane *Platanus orientalis* and the American sycamore *Platanus occidentalis*, in his nursery in Vauxhall, London, and he may have been the first to recognise this hybrid from a specimen growing there. I think it is fitting that such a popular London tree was the result of a chance natural hybridisation of two non-native trees from different parts of the world – an event that occurred in London itself. Yet the plane's Latin name, *P.* x *hispanica*, suggests that the London plane originated in Spain. We will probably never know...

LOCATION
High Street,
Buckden,
Saint Neots,
PE19 5TA

OS GRID REF
TL192677

One of the Buckden Towers Planes, the oldest London plane trees in Britain. This is the larger of the two trees, growing on a small wooded mound. Its impressive towering limbs dominate all around.

14. BURNHAM BEECHES

Beeches, not Beaches

Burnham Beeches is an ancient wooded common situated in Buckinghamshire. It covers an area of approximately 220 hectares (540 acres) consisting of woodland, wood pasture, wetland and heath. An exceptional place, it has been designated a Site of Special Scientific Interest (SSSI) since 1951, a National Nature Reserve since 1993 and a Special Area of Conservation since 2005. Acquired by the City of London in 1880, Burnham Beeches is now owned and managed by the authority, preserving it from developers and residential use. It was purchased under the Open Spaces Act of 1878, which sets out the City of London's duty to protect and conserve the common. Today it still provides a public open space for all to use and enjoy.

LOCATION
Lord Mayor's Drive, Slough, SL2 3PS

OS GRID REF
SU950850

Burnham Beeches, as its name suggests, is famous for its beech trees (*Fagus sylvatica*), particularly the ancient pollards with their wonderful arboreal shapes. Many are between 400 and 500 years old. It was a popular area to visit in Victorian times as it was only 25 miles from St Paul's in the centre of London. The Victorians gave names to many of the pollards. One named 'Jumbo', sadly long gone, was illustrated in *The Graphic* newspaper in October 1883. A more recently named tree, the Cage Pollard, was chosen as one of the 50 Great British Trees to honour the Queen's Golden Jubilee in 2002. It is now heavily propped with a mixture of timber and tubular metal props. The structure holds its hollow trunk and limbs in place, aiming to slow down the tree's inevitable collapse and increase its longevity into the future.

Fame has brought problems to the Cage Pollard. It had to be fenced off after featuring in the burial scene of the film *Robin Hood: Prince of Thieves* (1991), when hordes of visitors flocked to visit the tree and to be photographed with it. This extra footfall resulted in compaction to the soil around the tree; it caused stress by reducing pore spaces available for air and water. Burnham Beeches has also been used as a film location for *Harry Potter and the Order of the Phoenix* (2007) and *Harry Potter and the Deathly Hallows* (Part 1) (2010).

Burnham Beeches historically would have been wood pasture – land the local people would have had the right to use. The trees would have been regularly pollarded and cut on a rotation system every 12 to 15 years to produce timber for firewood. Domestic cattle and sheep would have grazed between the trees, keeping areas of pasture open, and in the autumn pigs would have been put out to root around, feeding

Burnham Beeches has many ancient pollards that have become 'natural sculptures'.

on beech mast and acorns. The Domesday Book of 1086 describes Burnham parish as having 'woodland enough to feed 600 swine'.

The regular pollarding of the trees stopped around 200 years ago. By the middle of the twentieth century grazing had stopped too. Left unmanaged and uncut, the pollards can become very top-heavy. The old trunks, many hollow and some very fragile, are no longer able to support the weight of their now large limbs. This has caused many of the ancient beech trees and some of the veteran oaks to break up, making them unstable and at risk of being blown down during gales.

There are around 300 ancient beech pollards at Burnham Beeches and fewer than 100 oak pollards (mostly *Quercus petraea*). There are also several standing dead trees. These trees (dead, dying and decaying) provide invaluable habitats for many rare and endangered communities of saproxylic invertebrates – creatures that depend on dead and decaying wood for at least part of their life cycle. Surveys have shown that Burnham Beeches is one of the richest sites for saproxylic invertebrates in Britain.

In the late 1980s restoration work on the common began. Scrub, young birch and holly were cleared from around many of the ancient trees, particularly the pollards. An ongoing plan to reduce the weight of overgrown limbs over several years was also put in place at this time. This will enable old beech and oak pollards to return to managed pollards, helping them to survive well into the future. Over 1,000 new pollards have now been created, to ensure continuity for future decades.

The management of the forest has also seen grazing animals return. From spring to autumn, traditional British breeds of cattle, such as the British white and Exmoor ponies, help to keep areas around the old pollards and throughout the woodland open. Pigs are also used to control bracken, otherwise very invasive. This grazing creates a more diverse range of habitats, reducing competition from new trees and scrub and returning the land to original wood pasture.

Wandering along the formal and informal paths (many with hard surfaces for year-round use) takes you through a vast range of habitats. As well as ancient woodland and amazing trees you encounter wood pasture, coppice, open heathland, grassland and wetlands. A late Bronze Age hillfort stands on Seven Ways Plain, on the opposite side of Lord Mayors Drive to the Druids Oak. This tree, thought to be around 800 years old, is the oldest in Burnham Beeches. It also has the largest girth, reaching around 9 m (29 feet).

New and old props being used to stabilise Burnham Beeches' most famous ancient beech, the Caged Pollard.

An article in *The Illustrated Sporting and Dramatic News* of 12 July 1879 vividly describes the setting:

> *Although one of the most picturesque and beautiful tracts of wild forest scenery in the country, it remained until the beginning of the present century a kind of terra incognita, a perfect sylvan solitude, its inhabitants partaking of the primitiveness of the scenery surrounding them, and preserving many ancient customs and rude manners peculiar to the locality.*

15. CAPON TREE

Forest survivor

Also known as the Jedburgh Oak is an 'English' oak (*Quercus robur*) – in this case actually a Scottish oak. It is one of the last surviving oaks of the ancient Jed Forest that once cloaked Scotland's border counties.

At around 800 years old, this valiant tree is certainly showing its age. Completely hollow, the wide, flared trunk has one particularly long, low branch that is supported along its whole length by wooden poles. Without these the whole tree would be resting on the ground. The main part of the trunk is also supported, which will hopefully keep the tree upright and flourishing for many years to come.

This ancient oak has certainly seen history. It grows along the forest edge next to one of the main routes into the border town of Jedburgh, where many people and animals would have travelled past it. It must have been a well-known landmark for centuries.

Why the oak became known as the Capon Tree is unclear, but it may have been from the Capuchin monks who sheltered under its branches on their way to Jedburgh Abbey.

Today the tree is used as a resting point during the annual Jethart Callant's Festival, inaugurated in 1947 and held every July. Jethart is the old name for Jedburgh and the festival commemorates the Scots' defeat of the English in 1317. The festival is led by the Callant (a young man), supported by his right- and left-hand men. Each is armed with a replica of the Jeddart staff, a long-bladed weapon.

A song sung during the festival is 'Brave Lads O' Jethart'. Its chorus celebrates the wood that may once have been part of the ancient Jed Forest:

> *Then here's to Border hill and wood*
> *And plain and bank and brae,*
> *To Border lads who boldly stood*
> *And kept our foes at bay.*

LOCATION
A68,
Hundalee Mill,
Jedburgh,
Scottish Borders,
TD8 6PE

OS GRID REF
NT 651188

An iconic survivor of the Jed Forest on the border between Scotland and England, this venerable oak is hollow and propped.

16. CHIRK OAK

'The Oak at the Gate of the Dead'

This large pedunculate or English oak (*Quercus robur*) is believed to be between 800 and 900 years old. It grows in the bottom of a valley, close to a road (the B4500) at Castle Mill, around 300 m (985 feet) below Chirk Castle. Access from the castle is via a very steep, sometimes muddy and slippery slope in a small boggy field.

When the oak was in one piece, its girth measured around 10 m (33 feet). Unfortunately, during a cold spell in February 2010 it spilt in two; one half of the tree has stayed upright, and the other half now lies on the ground.

The half now lying on the ground survived for a while, but is sadly now completely dead. But although the tree may have died in one sense, in another it is very much alive. Ancient oak timber, both dead and alive, is home to many other things, including rare invertebrates able to exist only in this ancient wood.

When I visited the Chirk Oak I saw a young birch growing directly out of this fallen dead section, enriched by the decaying wood. A blue tit was nesting in one of the oak's decayed hollows.

Growing on the bank above the named oak, just metres away, is another old oak. It is not quite as large in girth, being around 8 m (26 feet), but is still a very impressive tree, with gnarly old bark and burrs clothed in bright green moss.

The larger split oak is most commonly known as 'The Oak at the Gate of the Dead' or 'Derwen Adwy'r Meirwon' in Welsh. This name derives from the tree's association with the Battle of Crogen which took place in 1165. The battle was fought between the Welsh and the English, close to the spot where the oak grows on the border between the two countries. Here the English army, led by Henry II, was ambushed and defeated by the Welsh army of Owain Mawr, 'Owain the Great'. The ambush, according to local legend, took place in a gap in the ancient King Offa's Dyke, an earthwork believed to have been constructed in the eighth century by the Mercian king. The Dyke consists of a bank and ditch that runs for around 120 miles (193 km) roughly along the border between England and Wales. Some of the fallen Welsh were supposedly buried in this section of Offa's Dyke, close to the ancient oak, hence the area's eerie name of 'Gate of the Dead'.

LOCATION
Chirk,
Wrexham,
LL14 5AF

OS GRID REF
SJ264377

Growing just below Chirk Castle, 'The Oak at the Gate of the Dead' (Derwen Adwy'r Meirwon) is now in two pieces, but still very much alive.

17. CLUMBER PARK – LIME TREE AVENUE

Double delight

Europe's longest lime tree avenue (*Tilia* x *europaea*) is around 3.5 km (a little over 2 miles) in length. It has 1,296 common limes planted along what is also the longest double avenue of lime trees in Europe. The trees are a naturally occurring hybrid between the small-leaved lime (*Tilia cordata*) and the large-leaved lime (*Tilia platyphyllos*); it was frequently used for avenues in the nineteenth century, particularly to create formal entrances to grand houses. This impressive avenue was developed by the 5th Duke of Newcastle in 1840. It was not designed to be a tree-lined entrance to Clumber House, the grand mansion demolished in 1938, but as a show of grandeur between the entrances to the enclosed estate.

Lime Tree Avenue was closed for a time during the Second World War, when it was used as open-air storage for munitions in hundreds of corrugated iron stores. The road between the limes was re-opened in 1955, after the tons of stored ammunition had been removed.

Surprisingly, the black bands still clearly visible on each of the trunks along the avenue are from bands of grease. These were painted on the limes in 1906 when the trees were found to be under attack from insects. Painted grease bands served to prevent the female winter moths climbing up the trunks and laying their eggs in the canopies. An old black and white image of the avenue, taken in 1955, shows the bands on the trees more clearly.

Similarly, white bands painted on lime street trees can also often still be seen today. They were painted to make the tree more conspicuous at ground level during the blackouts of the Second World War.

LOCATION
Clumber Park,
Worksop,
Nottinghamshire,
S80 3AE

OS GRID REF
SK626751

A section of the remarkable double avenue of limes at Clumber Park, over 2 miles (3.5 km) in length. Some trees still show the marks of grease bands painted on them in 1906 as protection from moths.

18. CLUNY REDWOODS

Nature's skyscrapers

It is very fitting that some of the largest giant redwoods in Britain are to be found in Scotland, particularly in the Scottish county of Perth & Kinross. For here the giant redwood (*Sequoiadendron giganteum*), also known as Wellingtonia, was first introduced as seed into the UK in August 1853. John Matthew, from Perthshire, encountered the trees – long known to and valued by Native American peoples – while gold prospecting in Calaveras County, California. He collected samples of seed and sent them back to his father Patrick Matthew, who duly propagated them.

LOCATION
Cluny House Gardens, Aberfeldy, PH15 2JT

OS GRID REF
NN878513

In a selection of extracts from a letter John Matthew sent to his father, dated July 1853 and printed in *The Gardener's Chronicle and Agricultural Gazette* in 1854, he describes meeting these giants:

> *Last Saturday I went along with one of my partners to see the 'Big Tree' discovered in Calaveras County near the head waters of the Stanislaus River .... we reached a closely-wooded bottom where the trees were more luxuriant than I have seen in any other region, with a good deal of under-brush through which we had to force our way, and in the swamp we found the Wellingtonia, whose dimensions are as follows:- Diameter at the ground 34 feet; diameter 20 feet up, 20 feet; diameter about 120 feet up, 14 feet; height, 290 feet; age estimated at nearly 3,000 years. I do not think, however, that it is so old, as I find that there is on average about 15 annual rings in the cross section of its wood near the root to the inch; and taking the tree above the swell of the root at 10 feet to the centre, that is 120 inches, at 15 years to the inch gives only 1,800 years. ... In the same swamp there are many other trees of nearly the same diameter. I stepped around several, 30 yards in circumference, while one which had fallen has a hollow inside fit to stable 50 horses.*

The two monumental giant redwoods in the woodland garden at Cluny House quite possibly date from this early introduction. The enormous, flared trunk on the larger of the two has a girth of over 11 m (36 feet) and is the largest recorded girth of any tree in Scotland.

Standing at the base of the tree it is impossible to see all of its 40 m (131 feet) plus trunk to the top. The second of the two redwoods a little further into the woodland garden can be only a few centimetres smaller.

A giant among giants. Scotland is home to many huge conifers, but one of the Cluny Redwoods has the largest trunk of any growing in Britain.

19. COWDRAY PARK CHESTNUT

A splendid survivor

The Cowdray Chestnut is a tree of immense size, both in girth and in height. In overall size it is the largest sweet chestnut (*Castanea sativa*) in England.

Also known as the Cowdray Colossus, this enormous tree has a girth measuring almost 12 m (39 feet) and is 26 m (85 feet) tall. There are many estimates of the tree's age, but it is generally thought to be somewhere between 300 and 400 years old. It has a huge crown which breaks quite low down, producing four main upright limbs. Each individual limb would make a substantial tree in its own right. Also visible are the splintered remains of its lowest limb; this would once have been a huge limb growing out horizontally not far above the tree's base, before presumably being lost in a storm. Other storm damage is also evident; the chestnut has lost branches, and has the look of a tree that has definitely gone through the wars. For all that, however, it has a very healthy canopy.

Below the crowded low limbs, the base of the trunk flares out to over 15 m (49 feet), adding to its enormous visual impact. The tree was fenced in 2019 with chestnut post and rail. This prevents ponies and cattle in the surrounding field from damaging the tree by chewing the bark around its base.

Many other large sweet chestnuts grow on the Cowdray Park estate. These include old remnants of avenues and individual trees, some with girths between 8 and 10 m (26 and 33 feet).

The Cowdray Estate has existed since the twelfth century. It covers around 6,677 hectares (16,500 acres), part of which was the old deer park in the area around Benbow Pond and Oaters Wood. In this area are found many of the estate's veteran and ancient trees, among them the historic Queen Elizabeth Oak (p 130-131) and other ancient oaks.
Also close to Benbow Pond is the John Cowdray Arboretum. Named after the 3rd Viscount Cowdray and with an area of 6 hectares (15 acres), it opened in 2009. The arboretum has been designed with mown paths that radiate out from the centre, like the spokes on a bicycle wheel, and has lots of interesting trees.

LOCATION
Cowdray Park,
Benbow Pond
Carpark,
Midhurst,
Surrey,
GU29 0BB

OS GRID REF
SU914222

Battered by the elements and age, this colossal sweet chestnut is still an awesome tree, unrivalled in overall size.

20. CROFT CASTLE – CHESTNUT AVENUE

Armada chestnuts

A few Spanish chestnuts (*Castanea sativa*) indicate the remains of the ancient chestnut avenue in the parkland of Croft Castle. Some are now completely dead, victims of the fungal disease *Phytophthora cinnamomi*. Yet both the dead and the living trees are beautiful, naturally sculpted by centuries of growing along what was the formal approach to the castle.

It is said that the chestnuts were taken from captured ships of the Spanish Armada fleet that sailed against England during the reign of Elizabeth I. They were then planted in the grounds of Croft in a pattern representing the battle formation of the Armada's fleet.

The Chestnut Avenue is almost 1 km (0.62 of a mile) in length. It starts out as a single avenue, while the middle section is a triple avenue. The final part of the avenue, now closed off to the public for safety, is a double row that now resembles a small woodland rather than an avenue.

Each of the old chestnuts is different and unique. Most have been pollarded many times over the centuries and are now a mixture of sun-bleached deadwood and healthy growth. Their branches twist and turn above their trunks like the Medusa's hair or the tentacles of a giant octopus. Their thick, deeply fissured bark, with the characteristic spiralling pattern so often seen on old chestnuts, echoes the swirling dress of a Spanish flamenco dancer.

An ancient tree trail guides you along the avenue. Part way along the trail you will find the ancient Quarry Oak, over 1,000 years old. It is one of the oldest trees at Croft.

LOCATION
Croft Castle
and parkland,
near Leominster,
Herefordshire,
HR6 0BL

OS GRID REF
SO45119

The avenue of ancient sweet chestnuts at Croft Castle is said to have been grown from chestnuts from captured ships of the Spanish Armada.

21. DARK HEDGES – BEECH AVENUE

Fantasy avenue

This stunning iconic beech avenue was planted in the eighteenth century by the Stuart family as a grand entrance to Gracehill House, their newly built Georgian mansion. (The house still exists as a golf club and restaurant.) They now form a tunnel just over half a mile (900 m) in length. It is composed of beautiful, silvered-trunked mature trees, known as the Dark Hedges.

The avenue was originally planted with around 150 trees, but there have been many losses over the last 200 years. Today less than 90 beeches remain, but these are still an impressive sight. Planted fairly close together along what is not much more than a country road, the veterans' proximity to each other hides many of the gaps that have opened up.

This atmospheric avenue of trees is well known to viewers of *Game of Thrones*. In the American cult fantasy series it features as King's Road, a route across the fictional land of Westeros. Today the site attracts visitors from across the globe and has become one of the most photographed locations in Northern Ireland.

A tree blown over during Storm Gertrude in 2016 has had part of its trunk carved into a themed door. This door, now in the restaurant at Gracehill House, is called door #7. It depicts a three-eyed raven and is one of a series of ten intricately carved *Game of Thrones* doors in Northern Ireland.

Beech trees are not among the longest lived of trees; somewhere between 200 and 300 years old is a good age for them. They tend to grow tall, up to 40 m (131 feet), and have wide canopies, making them particularly susceptible to damage in storms and high winds, especially as they age. Beech trees can live much longer when pollarded. This is usually done from a young age and continued at regular intervals throughout the tree's life.

This majestic avenue is thus definitely at the veteran stage of its life, but will hopefully survive for many seasons to come. The Bregagh Road along which the trees grow was closed to traffic in 2017, in an attempt to protect the trees further.

Free signed parking is close by, requiring only a short walk to the Dark Hedges. There is thus no need to park along this amazing avenue, which risks damaging these venerable beech trees.

LOCATION
Bregagh Rd,
Stranocum,
Ballymoney
County Antrim,
BT53 8PX

OS GRID REF
NW212906

This magical avenue of beech trees looks particularly good when spring sunshine highlights the silvered trunks.

22. DRUMMOND CASTLE – BEECH AVENUE

Beautiful beeches

The entrance to Drummond Castle is an extremely impressive one. Passing on to the driveway through wrought iron gates dating from the eighteenth century, it looks like a narrow country lane lined with trees. This avenue of European beech trees (*Fagus sylvatica*), an undulating, single-track road to the castle and gardens, starts to feel quite imposing; the towering, silver-trunked trees closely flank visitors as they walk or drive between them for more than a mile. More than 250 of the mainly mature beech trees remain, with a few younger trees replacing those that have been lost over the years.

The avenue is thought to have been started by the 4th Earl of Drummond in the late seventeenth century. It was then apparently composed of four rows of trees that led from the castle all the way to Perth, a distance of 32 km (20 miles). Most of the avenue today was planted in the late eighteenth century, however, making some of the original trees around 200 years old.

European beech trees are sensitive to drought. In the UK the species seems to be slowly retreating north, as the south is becoming too dry and warm for them; they prefer cooler temperatures in which they thrive and do well. Scotland, generally cooler and wetter than the south of the UK, has more favourable environmental conditions. This avenue of beeches, generally healthy and happy, bears testament to this.

Although beeches make up the majority of trees along this amazing entrance driveway, they do not extend along the avenue's entire length. Around two-thirds of the way along lime trees (*Tilia* x *europaea*) and other mixed broadleaf trees also become part of the long drive.

This stunning driveway has to be one of the most beautiful and picturesque in all of Scotland. Fresh, lime-green leaves in spring turn darker green through summer, allowing the skies to cast moving shadows though their dancing forms. In autumn they turn glorious autumnal coppery shades before falling to reveal towering, branched skeletal forms.

LOCATION
Muthill,
Crieff,
Perthshire,
PH7 4HZ

OS GRID REF
NN837179

The amazing European beech avenue leading to Drummond Castle is over a mile (1.6 km) in length. It still features about 250 of the original imposing trees, now around 200 years old, as well as some younger replacements.

23. EPPING FOREST

London's largest open space

Epping Forest is another of our great royal hunting forests, designated as such in the twelfth century by Henry I. Like many ancient forests in Britain, Epping is associated with a mix of rare and important habitats, from areas of heathland and grassland to wetlands and wood pasture. It is classed as semi-natural ancient woodland as, although the forest trees may initially have occurred naturally, the forest has been managed for centuries through the coppicing and pollarding of the trees. Such skilful management has made Epping the special place it has become.

Epping Forest is the largest open space in the London area. Straddling the border between London and Essex, it covers over 2,500 hectares (6,185 acres) and was once part of a far larger woodland. Under the Norman kings it became the Forest of Essex and by the twelfth century covered much of the county's wooded areas. In the thirteenth century this vast forest was broken up into smaller woodlands, one of which became Waltham Forest. This was later broken up further, becoming the smaller forests of Epping and Hatfield.

Epping is one the few remaining large-scale examples of ancient wood pasture in lowland Britain. It contains close to 55,000 veteran pollarded trees, including 85 per cent of the UK's veteran and ancient beech pollards. This makes the forest home to Europe's largest single population of ancient beech pollards.

Several of the areas within Epping Forest have magnificent names, such as The Stubbles, Great Monk and Little Monk Woods, High Beach, Bury Wood and Barn Hoppitt. This last, close to Queen Elizabeth's Hunting Lodge, is an area of 71 hectares (175 acres); it includes an ancient pollard wood, some of whose oaks are around 500 years old. This is a fine example of oak wood pasture, providing lots of open areas among the trees for grazing. Here the trees offered shade to livestock and deer grazing on the more open area of Whitehall Plain.

An area known as Loughton Camp, located almost centrally in the forest, is the site of an Iron Age earthwork from around 500 BCE (a second ancient earthwork, Ambresbury Banks, lies within the forest at its northern end). As a result the area is made up of ancient mounds with forest tracks that run over and between the two sites. A sense of mystery envelops you as you explore the different levels of the beech woodland that surrounds the earthworks.

In the eighteenth century Epping Forest was the haunt of the infamous highwayman Dick Turpin, with the area around Loughton

LOCATION
The Warren,
Loughton,
Essex,
IG10 4RW

OS GRID REF
TQ409956

This area of lapsed wood pasture at Epping Forest shows both ancient beech pollards and a huge lapsed coppiced beech at the back.

Camp believed to be one of his hideouts. Unlike the romanticised figure of Robin Hood in Sherwood Forest, the legendary hero in folklore who robbed the rich to help the poor, Dick Turpin and his gangs were all known criminals. They were brutal thieves, involved with poaching, horse theft, robbery and murder.

Dick (Richard) Turpin's birthplace in 1705 was the Rose and Crown, now the Bluebell Inn, a public house in Hempstead, Essex. Opposite the pub a group of pollarded veteran oaks once stood around an area known as Turpin's Ring and used for cock fighting. Sadly these oaks no longer exist.

As well as the veteran beech and ancient beech pollards that make up most of the older trees in the forest, we also find areas of oak and hornbeam pollards, while lapsed beech coppice now supports growth many metres in height.

The beech has become the dominant tree species, being able to reproduce and grow faster than the oak and hornbeam. In many cases its dominance has shaded them out and has caused the decline of many other species of pollards.

The management of wood harvesting from these trees stopped in the mid-nineteenth century. At that time there have would have been the three types of wood pasture, or pasture woodland, managed in Britain: beech-oak, hornbeam-oak and mixed oak woodland, which contained both open grassland and wetland areas. Remnants of all these types of woodland still survive in the forest.

Some of the beech trees in the forest are truly ancient, while trees known as the 'Coppards' are certainly centuries old. Some estimates believe the Coppards to be 1,000 years old and they are beautifully, individually unique. The old, coppiced stools have had their coppiced stems pollarded, producing a hybrid between a coppice and a pollard.

Beech forest is the most colourful of our forests. The new spring foliage is vibrantly lime-green, with the leaves darkening though the summer before turning into the fiery shades of golds, reds and yellows of autumn. A beech woodland on a sunny autumn day, with golden-leaved paths and sun filtering through the colourful canopies, is a truly wonderous place.

An ancient oak pollard in Epping Forest. Like all pollards, its timber has had many uses, from animal fodder and fuel to material for construction.

The beautiful 'Coppards' of Epping Forest are the result of an unusual form of tree management. These coppiced trees have at some later stage also been pollarded and may be up to 1,000 years old.

Autumn is also the time of year when beech trees produce their fruits, the beechmast – a favourite food of pigs put out in the forest by the commoners during pannage. These, along with acorns on a good or 'mast year', would cover the forest floor.

In 1871–2 the City of London Corporation won a battle to save the forest from enclosure of forest land for development which was deemed to be unlawful. The Corporation has owned and managed the forest since 1878; it also owns a further 737 hectares (1,821 acres) of surrounding land as a protective buffer for the forest.

24. THE FLORENCE COURT LIME

Ireland's greatest broadleaf

The parkland surrounding the Palladian Georgian mansion at Florence Court contains some of the oldest silver firs in Ireland. It also features large cedars, a common feature in many eighteenth-century estates in Britain and Ireland.

The grounds, which cover an area of around 129 hectares (318 acres), are also home to one of the original fastigiate yews, known as the Irish or Florence Court Yews. These trees were found growing wild on the Cuilcagh Mountain above the Florence Court estate in the 1760s. George Willis, the tenant farmer who found them, was struck by the upright form of these yews, two of which he dug up. One he planted in his own garden and the other he gave to his landlord, Lord Enniskillen. The latter tree still grows on the estate and is the 'Mother' of all the Irish yews grown worldwide. One of the more spectacular trees in the parkland is an enormous common lime (*Tilia* x *europaea*). It may not be the tallest tree on the estate at just over 20 m (66 feet) tall, but its huge trunk of almost 11 m (36 feet) in girth makes this the largest girthed deciduous tree in Ireland. The lime grows close to the main entrance of Florence Court estate, along the path through the parkland to Cottage Wood. Further along the path is the woodland glade where the famous Irish yew grows.

The huge trunk of the Florence Court Lime is deeply fissured and knobbly; its main branches break from low down and have quite a wide spread. Much of its root plate is also visible, snaking out above ground around the base. Because the lime's short, deeply fissured trunk gives it an almost divided appearance, some experts believe that it is possibly a bundle planting. This practice, common in the eighteenth century, involves literally planting a bundle of sapling together in the same hole; over time it fuses together to produce a single tree. This creates a larger tree more quickly than using a single sapling.

The estate has been managed by the National Trust since 1953 and was declared an Area of Special Scientific Interest (ASSI) in 2012. Oak is the dominant tree species, complemented by ash, beech, hawthorn, horse chestnut and common lime trees. The parkland is also home to rare orchids, lichens and fungi, as well as many rare and endangered invertebrates.

LOCATION
Florence Court,
Enniskillen,
County Fermanagh,
BT92 1BZ

OS GRID REF
NV272002

Is the huge trunk of this common lime a result of a 'bundle planting'? We will probably never know. Either way it is truly impressive.

25. FREDVILLE PARK AVENUE

Remembrance of trees past

The remnants of a now lost avenue of sweet chestnuts (*Castanea sativa*) stretch across Fredville Park. They lead towards the site where the Plumptre family mansion once stood, through parkland now very different to what it was in previous centuries. In 1250 Fredville Manor existed on the site, since when a succession of different buildings, houses and families have left their mark on the estate and surrounding parkland.

Fredville Park covers 100 hectares (247 acres). It once had many more trees than it does today, but the survivors include many ancient and veteran oaks and chestnuts.

The Georgian mansion, developed by John Plumptre from a farmhouse in 1750, was mostly destroyed by a fire in 1942; its remains were demolished in 1945. Only the clock tower and some stable buildings, sited within an iron-fenced enclosure, provide evidence of the once grand mansion that stood there. Within this enclosure of around 5 hectares (2 acres) grows the Majesty Oak (p. 114), located roughly to the westside of the parkland.

The 1st edition of the Ordnance Survey map (1862–75) still features the remnants of the old avenue that ran from Fredville House into the park. Yet even in the nineteenth century trees were already clearly missing along the avenue, and very few remain today.

One of the surviving sweet chestnuts was famously known as the Step Tree. Postcards and photographs from the early decades of the twentieth century show the old chestnut with a flight of wooden steps and rails, leading up to a small platform or terrace. Here, it is said, the Plumptre family would take afternoon tea. It is also said that up to 20 people could fit on this raised platform. Having seen the tree, however, I think that would have been a bit of a squeeze!

Nevertheless, considering the age of the tree and its time as the Step Tree, this valiant chestnut has fared remarkably well. Although now propped, with some decayed and broken branches, it still retains three large limbs. These make up the tree's substantial and healthy canopy, producing a crop of chestnuts most years.

LOCATION
Fredville Park,
Nonington,
Kent,
CT15 4JG

OS GRID REF
TR256512

These ancient sweet chestnut trees are two survivors of a once much larger avenue that reached across Fredville Park.

26. GLEN LYON ASH

A splendid isolation

This beautiful pollarded ash (*Fraxinus excelsior*) is one of the oldest and certainly the largest in Scotland. It has an impressive girth of nearly 7 m (22 feet) and could be as old as 400 years.

The monumental ash sits on the sloping side of a field along a country road above the River Lyon, between the villages of Glen Lyon and Fortingall. Glen Lyon, reputed to be the longest enclosed glen in Scotland, was described by Sir Walter Scott as 'the longest, loneliest and loveliest glen in Scotland'. All alone with only sheep for company, the tree enjoys a truly peaceful and beautiful spot.

This ancient ash was probably once part of the far larger forest of Glen Lyon, much of which still exits. Its unmissable silvered, short and stumpy trunk, patterned by thousands of small furrows and covered in mosses and lichens, is partially covered by its large circular crown. The crown is healthy and made up of many tall, young, pollarded limbs.

There are many myths associated with ash trees, particularly involving snakes. One suggests that drawing a circle around a snake with an ash stick would trap the snake within the circle. The story may derive from an experiment reputedly carried out in the first century CE by Pliny the Elder, a celebrated Roman natural historian. Another myth claims that a branch from an ash tree could kill an adder with a single blow.

Close by is the Fortingall Yew, believed by many to be the oldest tree in the UK, or indeed in Europe. Its age is the subject of debate, but it is probably between 2,000 and 3,000 years old.

LOCATION
Camusvrachan,
Perth and Kinross.
Nearest postcode
PH15 2NL.

OS GRID REF
NN622478

The lovely Glen Lyon Ash, perhaps once part of a larger forest, now rests in splendid isolation at the edge of a field, with only sheep for company.

27. GLENMORE FOREST

Glorious 'grannies'

The high mountains that make up part of the Glenmore Forest Park lie within the heart of the larger Cairngorms National Park. The area is still a sanctuary for remnants of the once huge ancient Caledonian Forest, estimated to have covered around 1,500,000 hectares (3,706 acres). Although approximately only one per cent of this original forest now remains, as separated remnants, these remarkable fragments still have many areas of ancient pine and birch.

The oldest of these ancient Scots pines (*Pinus sylvestris*) are known as 'granny pines'. Many of the older trees may be up to, and possibly over, 400 years old; one individual in a remnant forest close by is known to be over 500 years old.

These ancient Scots pines grow in a wide range of forms, both in size and in shape. Many have large single trunks that grow to around 25 m (82 feet) in height, with wide spreading crowns. Others are much shorter. Some have huge, multi-stemmed trunks, deeply fissured, with the silver-grey or chestnut-brown bark associated with these amazing trees.

Looking up into their canopies, you can see how centuries of howling mountain winds and heavy falls of snow have all taken their toll; there is evidence of broken and missing limbs. But this all adds to the trees' character and shows their steadfastness in the challenging environment they have chosen as home.

Some of the best of these ancient pines in the Glenmore Forest grow along a waymarked route that takes you to the Pass of Ryvoan. Here a well-used trail goes past *An Lochan Uaine* (The Green Lochan), and steep, tree-covered hillsides. It continues through mixed evergreen and deciduous forest with sculptural junipers, heathers and wild bilberries, all carpeting the ground beneath beautiful old birches (known as Birks in Scotland) – both *Betula pendula* and *Betula pubescens*, with their lichen-clothed branches.

The Cairngorms in the Scottish Highlands is home to almost half of all of the ancient Caledonian Forest in Scotland. It is the most extensive total area of native pinewood in the UK.

Pinus sylvestris is Scotland's national tree, and the only truly native pine in the UK. However, it also has the widest distribution of any conifer in the world, ranging from Scandinavia in the north down to southern Spain.

LOCATION
Glenmore Visitor Centre, Cairngorm Road, Glenmore, Aviemore, PH22 1QU

OS GRID REF
NH977097

A huge, multi-stemmed 'granny pine' high in the mountains of Glenmore Forest.

28. HETHEL OLD THORN

Rooted in a rebellious past

Amazingly, what is probably one of the oldest hawthorns (*Crataegus monogyna*) in Britain, if not the oldest, is protected by one of the UK's smallest nature reserves. This tiny reserve, covering 0.025 of a hectare (0.061 of an acre), contains just one ancient hawthorn, some herbage and an interpretation board; it is surrounded by a wooden and wire fence for the old thorn's protection. The area was given to the Norfolk Naturalists Trust in 1960 by the late Mr F W Myhill, a local businessman. It is now managed by the Norfolk Wildlife Trust.

The Hethel Old Thorn is also known as King John's Thorn, as rebel peasants are said to have gathered there during the king's reign in the thirteenth century. If true, this would make the hawthorn at least 800 years old.

In his *Register of Remarkable Trees*, published in 1841, the botanist James Grigor notes that 'in looking upon it, one would suppose it had been here for thousands of years'. The thorn would have probably been quite a large tree at that time; it had a canopy spreading 31 yards (28.3 m).

In 1755 the girth of the Hethel Old Thorn was measured at just over 9 feet (2.75 m). By 1841 its girth had grown to 12 feet 1 inch (3.68 m).

Through decay the old hawthorn has shrunk somewhat, both in trunk and canopy, over the years. Its appearance now is very different from the days when it was portrayed in old illustrations sheltering livestock and people, with wooden props holding up many of its branches. But it is still a wonderful old tree, each year producing a good crop of fruit (haws) that attracts many birds as a winter food source.

Hethel Old Thorn can be accessed about 300 m (984 feet) along a public footpath that runs alongside the picturesque All Saints Church.

LOCATION
All Saints Church,
Church Lane,
Bracon Ash,
South Norfolk,
NR14 8HE

OS GRID REF
TG171005

Not looking very remarkable as ancient trees go... Yet this valiant hawthorn, one of the oldest in Britain, may have been a meeting place since the thirteenth century.

29. HOLKER GREAT LIME

A giant among trees

The Holker Great Lime is huge, enormous, immense: there are simply not enough superlatives to prepare you for the size of this tree's massive trunk. The spectacularly deep fluting of the trunk is more akin to the buttresses of an evergreen tree from the tropics than a deciduous tree of the British Isles.

In spring its fresh green leaves draw you in and under the arching low branches, covered with a mass of epicormic growth above the wide, squat trunk. You approach the tree through a magnificent pathway, carpeted by thousands of white, wild garlic flowers on each side.

The common lime (*Tilia* x *europaea*) is a natural hybrid between the large- and small-leaved limes (*Tilia platyphyllos* and *Tilia cordata*). The Holker Great Lime has the largest girth of any common lime in the UK, reaching an impressive 8 m (26 feet), and is around 22 m (72 feet) in height.

This giant tree is growing in the 10-hectare (25-acre) gardens of Holker Hall. Owned by the Cavendish family, house and garden have passed through the family line since the early sixteenth century. The magnificent lime is thought to be of seventeenth-century origin. It was probably planted as part of a formal garden, surrounded by 494 hectares (200 acres) of parkland.

It is truly one of our great British ancient trees.

LOCATION
Holker Hall Gardens,
Cark-in-Cartmel,
near Grange-over-Sands,
Cumbria,
LA11 7PL

OS GRID REF
SD359773

The deeply fluted trunk of the Holker Great Lime is immense and appears to be very healthy.

30. KEW BLACK LOCUST

One of 'Kew's Old Lions'

The original arboretum in the Royal Botanic Gardens, Kew dates from 1759. It was approximately an area of 5 acres (2 hectares), set within the 9-acre (3.6-hectare) botanic garden and pleasure grounds of Princess Augusta. In 1762 many rare trees and shrubs were added to the arboretum by Lord Bute, who acquired them from his late uncle the Duke of Argyll, after his death. The Duke of Argyll's estate was just a few miles from Kew, in Whitton.

One of the trees from the 1762 donation that still survives today is a beautiful, stumpy and gnarled ancient black locust (*Robinia pseudoacacia*). The species, from the south-eastern US, was encountered by British colonists in 1607, and was subsequently introduced to Britain around 1640. Kew's specimen is now the oldest growing in the UK with a verified age. Its girth measures 5.40 m (17 feet 4 inches) while its deeply furrowed trunk has almost as much dead wood as living – a powerful testament to its age.

A series of three metal bands, fitted to the tree in the 1960s, were tailor-made to help keep the trunk from breaking apart. These have long ago stretched beyond their limit, with the tree's still expanding waistline causing them to snap. The tree has also been fitted with the modern 'cobra' support system, in which polypropylene cables are attached between the limbs in the canopy. This enables the limbs to move and flex, but gives them added support.

The tree's crown is now made up of just a few main branches. This adds to its appeal, giving it a very oriental appearance. This is particularly evident in spring, when it produces a wonderful head of lime-green leaves, followed by long trailing strings of flowers, loved by bees.

Another surviving tree at Kew believed to have come from Lord Bute in 1762 is an oriental plane (*Platanus orientalis*) close to the Orangery. Yet another is the iconic pagoda tree (*Styphnolobium japonicum*), now heavily propped with metal supports and brickwork. The ginkgo (*Ginkgo biloba*) growing close by may have been one of the earliest planted in Britain, planted in the original arboretum in 1762. These trees are known as 'Kew's Old Lions'.

However, the oldest tree in Kew Gardens is a sweet chestnut (*Castanea sativa*) from the late 1690s – even older than the Gardens themselves. The sweet chestnut grows near the Mediterranean Garden; it may have been an original boundary tree or formed part of an avenue.

LOCATION
Royal Botanic Gardens,
Kew,
Richmond,
Surrey,
TW9 3AE

OS GRID REF
TQ187776

This old *Robinia* is at its showiest with yellowing autumn leaves.

31. KING OAK

Royal remnant

The oak woodland that survives in the Midlands region of Ireland, in which Charleville Castle sits, is thought to have been the meeting grounds of Irish Druids. It is the remnant of the ancient oak forest that was a common part of the Irish landscape centuries ago.

The King Oak, a pedunculate or English oak (*Quercus robur*), is thought to be around 400 years old. It has the most amazing shape, in profile resembling a wide-open fan. The lowest branches start just above ground level, then extend out horizontally at right angles to 27 m (89 feet) from the wide, flared trunk. The lower, moss-covered limbs are propped and supported with wooden posts to prevent them from breaking away from the tree. There is an added incentive here because a legend associated with the tree claims that should any of its branches snap and break off, a member of the Howard-Bury family will die.

Lieutenant Colonel Howard-Bury, the last of the family to own Charleville Castle and estate, died in his early eighties in 1963. This was the same year in which the King Oak was struck by lightning, creating a long scar from the top of the tree all the way to its base.

It is also from Lt Col Howard-Bury's reconnaissance expedition in 1921, exploring routes to climb Mount Everest, that the legend of the 'Abominable Snowman' or 'Yeti' derives. He described seeing human-like figures and tracks at an altitude of around 20,000 feet (c.6,000 m), which his Sherpa guides explained as the tracks of the *Methokangmi*, a legendary snowman. Howard-Bury himself thought they were most likely the tracks of grey wolves.

The spread of the tree is actually wider than its height, which is around 20 m (66 feet). Throughout its canopy it is furnished evenly with wide, spreading branches. There are many other great oaks on the estate, some with larger girths, but the overall size and shape of this fine tree justify its title as King Oak.

LOCATION
Charleville Castle,
Tullamore,
Co Offaly

Tall and wide, this tree certainly deserves its name of the King Oak.

32. KINGLEY VALE YEWS

Magical forest

To see yews growing in their natural state, there is no better place to visit than the enchanting yew forest at Kingley Vale. This rare community of yews covers much of the southern chalk slopes of this National Nature Reserve, set within the larger Site of Special Scientific Interest (SSSI) in West Sussex. At just over 200 hectares (494 acres), it is one of the largest and finest yew forests in Western Europe.

It is amazing that this truly impressive forest with its ancient yews survives at all, as during the fourteenth century yew wood was prized for making longbows, used for both hunting and for warfare. Longbows are famed as the decisive weapon in the Hundred Years' War (1337–1453).

The forest at Kingley Vale is predominately made up of yew, accompanied by ash, hawthorn, holly and oak. Sadly, much of the ash has been infected with ash dieback (*Hymenoscyphus fraxineus*) in recent years and most is unlikely to survive.

Yews of various ages are scattered throughout the forest. However, there are areas in the lower part of the valley where the older, ancient and veteran yews are grouped, almost like old families that have stayed together. Here in these groves are found yews with huge trunks of all shapes and sizes, contorted over the centuries. Many have the characteristic deeply fluted trunks often associated with ancient yews. Some have big, bulging bellies and others completely hollow shells: each is unique.

Many of these naturally growing trees have twisting branches. They snake out from the parent tree, reaching down to the ground and producing arches and alley ways, through which you have to scramble to reach cave and cathedral-like spaces. Some branches reach the ground and layer themselves, producing new roots, re-growing as siblings around their parent and increasing the forest's spread.

These are known as walking trees. I love the thought of trees being able to walk slowly further and further from their parent, exploring and extending their range by continually rooting the lower outstretched limbs. This is something you will only see in a natural forest, a place in which trees can develop without human intervention, in contrast to the equally amazing huge ancient yews of the churchyards, whose lower limbs have all been removed to around head height and above for easy access.

The ground beneath and between the groves of these marvellous old yews is barren and dark, with no vegetation able to grow in their shade.

LOCATION
Kingley Vale National Nature Reserve,
Chichester,
Sussex,
PO18 9BE

OS GRID REF
SU824088

The forest yews of Kingley Vale all assume unique shapes and forms, typical of naturally grown trees.

However, the ground is softly cushioned with decades of accumulated leaf litter. The absence of other plants, the quietness, the sunshine that is filtered (at best), all add to the sense of being somewhere magical, unworldly and eerie.

Unlike ancient oak woodland, where invertebrates have evolved along with the oaks over hundreds of years, to live, feed and breed on and within the decaying timber, yews – even though they are as every bit as old, and in some cases much older than our oldest oaks – have very little that lives along with them. The fruits with their fleshy red seed covering (arils) are eaten by birds. (The seeds themselves, poisonous to most birds, are eaten by the non-native ring-necked parakeets that have naturalised in the south of England.) Yet apart from a few caterpillar species, not much else can tolerate the toxins present in much of the yew.

According to one local legend, these yews were planted as a memorial to a battle fought at this site in CE 859 between the Anglo Saxons and the Vikings. Another legend says that the forest is haunted by Vikings from the battle or is a meeting place for Druids. Yet another claims that the trees transform into human figures on moonlit nights.

Some of these amazing old yews are thought to be 500 years old, and could in fact be a lot older. As a result, in 1952 Kingley Vale was designated one of the first National Nature Reserves in the country – due in part to the efforts of twentieth-century ecologist Sir Arthur Tansley. He campaigned to have the special habitat that makes this unique forest protected. A memorial stone in his honour, known as the 'Tansley Stone', is sited near the top of Bow Hill, the highest point above the yew forest with far-reaching views across the South Downs National Park. The stone has a metal plaque. Its inscription reads:

> *In the midst of his nature reserve which he brought into being this stone calls to memory Sir Arthur George Tansley, FRS who during a long lifetime strove with success to widen the knowledge to deepen the love and to safeguard the heritage of nature in the British Isles.*

The 160 hectares (395 acres) that make up the National Nature Reserve within the SSSI also include the Devil's Humps. This group of four circular Bronze Age barrows is set in an open area on the western side of the forest, between West Stoke and Bow Hill.

During the Second World War, Kingley Vale was used as a firing range by British and Canadian troops. Evidence of this can still be seen in the trees damaged by shrapnel. Some of the ancient yews that now stand as sun-bleached corpses were probably used for target practice.

Today the woodland is managed by Natural England, which provides an information centre and nature trail.

These natural yew tunnels at Kingley Vale are formed by the lowest branches arching down to the ground.

33. KINGS INNS LONDON PLANE

The hungry tree

Just within the south gate that takes you into the grounds of Kings Inns, a law school on Constitution Hill which has public access, grows a remarkable tree.

This may not be one of the oldest London plane trees (*Platanus* x *hispanica*) in Ireland, but it is a truly remarkable one. Probably less than a century old, its appetite extends beyond the usual nutrients, water and sunlight required for growth. This tree has notoriously decided it needs some extra iron!

It has grown over and enveloped much of an old wrought iron park bench, probably older than the tree itself. It looks as if the tree is slowly devouring it.

This plane tree is listed as one of the heritage trees of Ireland by the Tree Council of Ireland, a recording normally reserved for trees of great age or rarity. The Kings Inn plane has become a local attraction because it is so unusual, however; it has even featured on the cover of a guidebook, *Secret Dublin*.

A tree preservation order was applied for a few years ago to give the tree extra protection, but this was refused: the site on which the tree grows is protected, and the authorities decided that this was enough. There was also consideration as to whether the bench, dating from the early nineteenth century, should also be listed for protection. This was refused in turn as it might have required removal of the tree to save the bench...

So the Kings Inns London Plane continues to delight its admirers – but do not rest on the bench for too long!

LOCATION
Kings Inns,
Henrietta Street,
Dublin 7,
DO1 KF59

OS GRID REF
SG158925

No, your eyes don't deceive you, this really is a bench and an over-friendly tree!

34. KINGSTON LACY – BEECH AVENUE

A gift for future generations

What an amazing birthday present.

This striking beech avenue lining what is now the B3082 was once the road that ran close to the entrance of William John Bankes's grand country mansion. It then passed the ancient monument of Badbury Rings, skirting the earth banks as it ran from Wimborne to Blandford.

These now veteran beech trees create a natural archway of regularly spaced silver-trunked sentinels, a dramatic 1.8 miles (3 km) in length. The trees are a stunning sight at any time of year: fresh green in spring, deep, rustling shade in summer, coppery-russet tones in autumn and leafless, naked forms in winter that glow starkly glow in the low sun.

An English politician and lifelong friend of the poet Lord Byron, Bankes inherited the estate of Kingston Lacy in 1835. He transformed the family mansion both inside and out, as well as planting the long avenue of mature beech trees. A gift to Bankes's mother, this has become one of Dorset's most famous landmarks and views.

A total of 731 beech trees originally made up the avenue. Of these, 365 were planted on one side of the road, one for each day of the year, and 366 on the other side, representing a leap year.

The new road was built as a turnpike or toll road (requiring travellers to pay a toll to pass through the gate and continue along the road). The money raised did in part pay for a new road, but it was also a source of revenue. Collected income was returned to the owner of the estate in which a road ran, in this case the Bankes family.

As you would expect, many of the beech trees have been lost over the years; survivors are now around 180 years old. Storms in both 1987 and 1990 hit the avenue badly, blowing over many trees on both occasions. In 2008 disease caused the felling of 21 trees which had become potentially dangerous.

The estate, including the avenue, is now managed by the National Trust. A new planting of hornbeams has been planted either side of the surviving original 1835 trees, continuing the life of this historic and spectacular avenue long into the future.

LOCATION
Blandford Road,
B3082, near
Wimborne,
Dorset,
BH21 4EA

OS GRID REF
ST972020

A small section of the avenue of beech trees at Kingston Lacy. This long avenue, 3 km (1.8 miles) in length, offers glorious colour for much of the year, while winter reveals the trees' striking naked forms.

35. MAJESTY OAK

A magnificent maiden

The Majesty Oak, also known by its older name of the Fredville Oak, grows in Fredville Park near Dover. A truly impressive tree, it now has a girth of more than 12 m (39 feet) and has reached 19 m (62 feet) in height. It is an English or pedunculate oak (*Quercus robur*) and the largest maiden oak in Britain.

An account of the oak in the 1790s appears in *The History and Topographical Survey of the County of Kent*, produced by the county historian Edward Hasted (1732–1812):

> *At a small distance from the front of Fredville-house, stands the remarkable large oak tree, usually known by the name of the Fredville oak. It measures twenty-seven feet round in the girth, and is about thirty feet in height; and though it must have existed for many centuries, yet it looks healthy and thriving, and has a most majestic and venerable appearance.*

Most other oaks of this girth have been pollarded at various times in their lives, increasing their girth 'un-naturally' (although oaks do naturally become stag-headed in old age) – but this does not necessarily mean that they are older. Pollarded oaks are usually shorter, stocky trees, but the Majesty Oak is taller and in generally good health for its age and size. It has lost some of its limbs, however, revealing hollow shell openings through which you can see daylight.

Other oaks known as Beauty and Stately once grew close to the Fredville Oak in the park. Drawings by Jacob George Strutt for his *Sylva Britannica* in 1824 show some of them, but it is not clear which the trees are. One that Strutt has named the Great Oak at Fredville does bear some similarity to the Majesty or Fredville Oak.

As well as the oaks in the park, the remains of an avenue of chestnuts used to run from Fredville House into the parkland. A few of these ancient chestnuts still survive today, no longer as an avenue but as remnants (p. 92).

LOCATION
Fredville Park,
Nonington,
Kent
CT15 4JG

OS GRID REF
TR256512

The Majesty Oak in Fredville Park, a massive maiden oak, reveals its completely hollow trunk, exposed by openings left by lost limbs.

36. MAPLEBECK YEW

Champion of the county

In the churchyard of the small, thirteenth-century church of St Radegund in the heart of Maplebeck grows the oldest yew tree in the county of Nottinghamshire.

The church and churchyard of this little rural village sit on a raised area above the tiny village green; the ancient yew grows on the east side above Church Lane. Both church and yew are mentioned in the *History, Gazetteer and Directory of Nottinghamshire* in 1864. The section on Maplebeck reads: 'the church is a neat fabric with a tower, surmounted by a short spire. In the church yard is a fine yew tree, supposed to be very old'.

The condition of the old yew today can no longer be described as fine. Most of it is very much alive, but it has also quite a lot of dead sections and has fragmented into two main parts, still joined at their base. Unfortunately, one of the fragmented sections on the north side of the yew is now completely dead. It is covered with the remnants of ivy that once grew on this fragment and probably caused its demise; the ivy is now dead too. There is also a lot of deadwood throughout the main living section. That said, this female yew has a girth of almost 7 m (23 feet) and is still an impressive sight in the village churchyard.

Interestingly, the dedication of the church to St Radegund is one of only a few in the UK. A German princess of the sixth century, she was seized by the invading Frankish king Chlothar I at a young age. Later, aged 18, she was compelled to marry him and become his queen. Chlothar was a violent man who arranged the murder of Radegund's brother. She escaped her murderous husband and in the early 550s founded a monastery in Poitiers, France.

LOCATION
St Radegund's Church, Church Lane, Maplebeck, Newark, NG22 0BS

OS GRID REF
SK711607

The fragmented shell of the ancient Maplebeck Yew still manages to maintain a healthy canopy.

37. MEIKLEOUR HEDGE – BEECH HEDGE

A Hedge above the Rest

The Meikleour beech hedge (*Fagus sylvatica*) has been officially recorded as the tallest hedge in the world since 1966. It towers over the A93, the main road between Perth and Blairgowrie, with an average height of over 30 m (c.100 feet). The hedge is tallest at its northern end, where it reaches around 36 m (118 feet) high. At approximately 530 m (580 yards) in length, it is truly an impressive sight.

The hedge was planted in 1745 by Jean Mercer of Meikleour and her husband Robert Murray Nairne. The Jacobite couple owned the nearby Meikleour House, for which the hedge acts as one of the boundaries at the eastern side of their estate. Robert Murray Nairne, one of the hedge's founders, also lost his life in the Jacobite defeat at the Battle of Culloden (16 April 1746). Sadly, he never saw his boundary hedge grow and develop.

Nearly three centuries later, the mighty Meikleour Hedge makes a truly spectacular sight all through the year. Driving along its length is particularly dramatic during the autumn, when it produces a wall of fiery, coppery tones.

Cutting the hedge is a huge task, carried out every ten years. During this time it is also remeasured. When the hedge was last cut a team of four took six weeks to get it into shape, despite using huge hydraulic lifts. I can only imagine what a dangerous and onerous job it must have been a century ago when ladders were used.

LOCATION
A93 Perth-Blairgowrie Road, Meikleour, Perth and Kinross, PH10 6QS

OS GRID REF
NO161389

Pruning a hedge one-third of a mile in length is a mammoth task. Meikleour Hedge is pictured here after its most recent trim in 2019.

38. MOTTISFONT PLANE

A lady of mystery

The grand plane tree (*Platanus* x *hispanica*) that grows close to Mottistone House in Hampshire may not be the oldest plane in the UK but at around 40 m (131 feet) it is probably the tallest, however, with a girth of 12 m (39 feet). Whether this is in fact a single tree has been the subject of much debate; it may be the result of a seventeenth-century practice known as bundle planting. This technique served to create a tree with a large spreading crown much more quickly than by using a single tree.

Another theory is that the Mottisfont Plane consists of two trees planted close together, which have now fused to make one large tree with enormous girth. Although it does have this appearance at first, so do many of the ancient plane trees from a similar period, which break low down and produce many large low limbs. Were they all planted in a similar fashion?

There are other possibilities to consider. The plane may have been cut back at an early age, or may have suffered some kind of environmental or physical damage caused by weather or an animal. For me the jury's out, though I lean towards it being a single tree. If so, that would make this plane the largest in the land. Whatever the answer, it is a truly awesome tree.

Known as London plane, *Platanus* x *hispanica* is a hybrid between *P. orientalis* and *P. occidentalis*. It is also known under the synonym *Platanus* x *acerfolia*. The hybrid may have originally been found growing in Vauxhall Gardens, London, by the seventeenth-century botanist John Tradescant (p. 64). It proved to be a popular tree for growing in the polluted city air, as its ability to shed its bark removed pollution from the trunk. London planes are still the most popular street tree planted in the capital.

LOCATION
Mottisfont Lane,
Mottisfont,
Romsey,
SO51 0LP

OS GRID REF
SU322468

The trunk of the huge Mottisfont Plane is thought by some to be an example of 'bundle planting' rather than a single tree. The jury's out...

39. MUCKROSS YEW

Hidden yew

On the shores of Lough Leane, within the Killarney National Park, are the remains of Muckross Abbey, founded in the fifteenth century as a Franciscan friary. Hidden within the vaulted cloister is a small, square courtyard in which a remarkable yew tree grows. It sits centrally within this space, slightly raised, its base circled by a ring of large stones. The yew's tall, knobbly trunk – deeply fissured, with spiralling red bark – reaches high up into the courtyard. Branches stretch out just below the roof line, letting the yew's canopy fill the space like a large, living umbrella.

The cloistered courtyard has a magical feel. Sunlight filtered through the yew's canopy creates delicate shadows that dance along the tree's furrowed limbs and trunk, then spiral around the lichen and moss-covered stones of the courtyard's arched walls.

The Muckross Yew is thought to be as old as the Abbey itself, making it more than 550 years old. It is the oldest in Killarney, and possibly in all of Ireland.

According to local tradition, the friary was built around the already established yew. However, its shape suggests that the tree has always been reaching for the light, and so has grown to such a height. Had the yew existed prior to the building, it is unlikely to have grown into such an unusual shape.

The yew's remarkable shape has always been a source of fascination. A quote from the *Dublin Penny Journal* in 1833 reveals its enduring appeal:

> *The stem of this remarkable tree, which there is no reason to doubt is coeval with the Abbey, is upwards of twelve feet in height, and about six feet six inches in circumference. It is preserved with religious veneration by the peasantry; and so awful is the effect produced on the mind of by its extraordinary canopy, that many persons shrink back with terror on entering within its precincts, and few can remain long without feeling an impatient desire to escape from its oppressive influence.*

The Abbey and yew are hidden away from the main town. Both are easily reached by a short walk from a signed car park, situated at the side of the N71.

LOCATION
Muckross Abbey,
Muckross Road,
Killarney,
Co Kerry.
Nearest postcode
V93 RR59.

The beautiful spiralling and twisting bark of this yew appears to have been drawn up by a vortex within the Abbey courtyard.

40. PLAS NEWYDD BEECH

Best in class

This absolutely huge and amazing beech (*Fagus sylvatica*), hundreds of years old, now sits hidden on a bank above a woodland path. You would hardly think it possible for a tree of this size to be hidden – but from the side facing the path, which is probably its widest side, the Plas Newydd Beech is almost completely obscured by undergrowth of scrub and young tree saplings.

Its multi-stemmed trunk measures around 11 m (36 feet) in girth and it is thought to have the largest trunk of any beech in Britain. This exceptionally large trunk is the result of the tree having been cut back or coppiced in its distant past, creating a deeply fluted, short appearance; it is only around 2 m (6½ feet) tall.

Many large limbs grow out from the short trunk. They reach upwards, branching out into a huge canopy and achieving a height of more than 25 m (82 feet). The beech's moss-covered roots also snake out above ground from around the huge base of the trunk, seeking to secure this monolith to its woodland bank.

The woodland in which the beech grows follows along the north bank of the Menai Strait. It is a beautiful spot, offering spectacular views across the Strait to the towering mountains of Snowdonia beyond. Within the woodland grow many other notable trees, including large maiden oaks and other large beeches. Some of these beeches were pollarded over a century ago.

In spring, this tree-lined woodland path leads you through bluebells, ferns and other wildflowers towards the rhododendron garden at its end. Dappled sunlight filters through the fresh, lime-green foliage of the beeches that dominate this woodland. The walk is peaceful and enchanting, with a magical tree at its heart.

The Plas Newydd estate, including the woodland, is now managed by the National Trust, to whom it was gifted in 1976. It was formerly the family home of the sixth Marquess of Anglesey.

LOCATION
Llanfairpwll,
Anglesey,
LL61 6DQ

OS GRID REF
SH525705

This enormous multi-stemmed beech at Plas Newydd has the largest girth of any beech in Britain. Its remarkable size is the result of coppicing several centuries ago.

41. POSSO SYCAMORES

Big is beautiful

The bigger of these two giant sycamores is the largest tree of its species (*Acer pseudoplatanus*) in Britain. Its girth – reaching an impressive almost 9 m (29 feet) is impressive enough, but the basal flare made up of its roots is astonishing.

The phrase 'the photo does not do it justice', often overused, is in this case fully justified. The tree's base is absolutely huge; you feel dwarfed sitting on it.

I first visited the Posso Sycamores 20 years ago, in 2001, and remember walking towards them across a small field behind the farmer's house. Even from a distance they are an impressive site. Closer up you can see their buttress-like roots, bigger than on any tree I had seen in the UK.

The tree nearer the farmhouse is the smaller of the two. It is by no means small, however, with a girth of 7.5 m (24 feet) and towering to a height of around 30 m (98 feet). Wide flaring roots around the base extend up into the statuesque bole, producing a deeply fluted, almost scaly trunk with an enormous canopy.

The larger of the Posso Sycamores is in the next small field, enclosed by a stone wall. It is one of my favourite trees. Its massive gnarly trunk and roots sprawl over a low bank, with the first whorl of branches formed low down, producing an incredibly wide crown. One of my favourite trees, it grows close to what remains of Posso Tower (now only part of a gable wall is left standing). These stone ruins belong to a tower house thought to have been built in the first half of the sixteenth century and occupied by the Naesmyth family, who possibly planted these two trees. They could never have imagined the legacy that their early planting has left. Later on Sir James Naesmyth, 2nd Baronet, another Naesmyth from Posso, became a Scottish botanist. He introduced the first larch trees into Britain in 1725.

These giant sycamores grow on private farmland, just to the west of Posso farmhouse. Permission must be sought from the owners before you visit them.

LOCATION
Posso Farm,
Posso,
Peebles,
Berwickshire,
EH45 9JW

OS GRID REF
NT202332

I first saw this tree more than 20 years ago in 2001. The larger of the two Posso Sycamores, it is still the most impressive sycamore I have ever seen.

42. PRESTON TWINS

Surviving twin

The English elm *Ulmus minor* was once a common sight in our treescape, being similar in numbers to the English oak. It grew in most parts of England and Wales, but was a particular feature of southern England. Huge trees growing to 30 m (98 feet) were commonplace, and elms appeared in many rural scenes of old masters' paintings.

During the 1970s the species was devastated by Dutch elm disease, a fungal disease (*Ophiostoma novo-ulmi*) spread by elm bark beetles. It is estimated that somewhere between 25 to 30 million trees died in this epidemic. The fungal disease was actually imported accidentally on logs from the US in the late 1960s; it did not come from the Netherlands, as the name might suggest.

Few mature elms now remain, but the species does survive as young suckering trees in countryside hedgerows (the beetles prefer trees between fifteen and twenty years old).

Brighton in southern England still has an abundance of mature elms, estimated to be around 17,000. This is partly due to geography: the area sits between the South Downs and the English Channel, both natural physical barriers. However, the mature elms' survival is also due to control measures put in place by the city council.

Preston Park on the outskirts of Brighton has many large elms. It is particularly well known for the Preston Twins, two huge old trees thought to be around 400 years old and believed to be the oldest and largest elms in Europe – and possibly the world! Both had hollow trunks measuring over 6 m (20 ft) in circumference; they used to stand side by side in the northern end of Preston Park's Coronation Garden.

Sadly, in 2019 one of the historically important twins had to be felled in order to save the other. It was discovered that it was suffering from the dreaded Dutch elm disease and was at risk of spreading it to its sibling. This tree had also suffered major damage in 2017 when it lost two limbs in a storm, tearing away a large section of its trunk. The crown of the surviving twin was reduced when the other was felled to avoid a similar fate.

Some of the timber from the felled twin has been used as 'deadeyes' – features of ship's rigging now fitted to the replica of the *Golden Hinde*. The original galleon, captained by Sir Francis Drake, circumnavigated the globe in the sixteenth century.

LOCATION
Coronation Garden,
Preston Park,
Preston Road,
Brighton,
BN1 6DS

OS GRID REF
TQ302062

The Preston Twins, two fine old elms in Preston Park near Brighton. The tree at the back left was badly affected with Dutch elm disease and had to be removed in 2019.

43. QUEEN ELIZABETH OAK

Shelter for a sovereign

It is amazing how many old oaks are named the Queen Elizabeth Oak, reflecting claims that they were visited by Elizabeth I. However, this incredible ancient oak in the old deer park on the Cowdray estate may have more justification for its name than most.

What is now the Cowdray Ruins was once an important early Tudor house, originally built in the sixteenth century. It is known to have been visited by Henry VIII and Elizabeth I before being destroyed in a fire in 1793.

The Queen Elizabeth Oak is the most famous of all the ancient trees in Cowdray Park. It is said that the queen sheltered from the rain under this tree in 1591 while hunting in the deer park. This oak, even four centuries ago, must have already been a substantial tree, ideal for providing shade, shelter and rest.

Now completely hollow, with a wide opening on one side, the Queen Elizabeth Oak has a squat appearance. This suggests that the tree has been pollarded at some time, or times, during its long life – one of a group of ancient pollarded sessile oaks (*Quercus petraea*) in the park. Its blackened and charred hollow interior shows signs of past vandalism, as a fire has burned the inside of the tree in its not-too-distant past. A post-and-rail fence has been erected around the oak for many years now to protect it from both people and animals, particularly cattle grazing in the surrounding field.

With a girth of 12.6 m (41 feet), the Queen Elizabeth Oak is the second largest sessile oak in England. The largest, known as the Marton Oak, is in a private garden in Cheshire. That particular tree is beyond hollow, its trunk having divided into three distinctly separate segments.

Close by is another ancient sessile oak, aptly named The Lady in Waiting as it is so close to the queen. This tree has a bulging waistline and a hollow that goes right through its middle. It too has been recently fenced for protection.

Both trees can be found close to the footpath by Steward's Pond, north of the small parking area beside Benbow Pond. A word of warning, however: the fields in this area can be very wet, particularly in autumn and winter.

LOCATION
Cowdray Park,
Midhurst,
West Sussex.
Nearest postcode GU29 0AY.

OS GRID REF
SU91406

From this side the Queen Elizabeth Oak at Cowdray Park appears to have an enormous and solid trunk. However, it is a completely hollow shell, with a wide opening on the opposite side.

44. ROSLIN CHESTNUTS

A Scottish sanctuary

Deep in the woods of Roslin Glen are two of Scotland's oldest and largest sweet chestnuts (*Castanea sativa*). Both have enjoyed this peaceful wooded sanctuary, in a valley above the North Esk river, for between 400 and 450 years. The spot was celebrated by Dorothy Wordsworth, the English poet and diarist and William Wordsworth's younger sister, who declared that she 'had never passed through a more delicious dell than the glen at Roslin'. It is the largest surviving stretch of ancient woodland in Midlothian.

The ancient Roslin Chestnuts are thought to date back to the time of Mary Queen of Scots. Although only metres apart, the trees are quite different in appearance to each other, one having fared much better than the other. They grow in an area of woodland that was once the grounds, and former site of, Rosebank House, built as a home for the Dowager Countess of Rosslyn.

Rosebank House was also the birthplace of Hector Macneill, an eighteenth-century poet and songwriter. The house was demolished in the 1950s. A group of seven old yew trees growing on a bank, possibly once planted along a boundary, still remain, along with other more exotic trees, as evidence of this old estate.

The first of the two chestnuts has a large swollen base and a huge, short trunk with characteristic twisting limbs. The second has a much shorter, stumpy trunk. Almost 8 m (25 feet) in circumference, it is covered in swollen burrs, while its large limbs, twisting skyward, resemble the huge antlers of some mythical stag. The trunks of both trees are clothed in silver and green lichens and mosses. Ferns flourish in their high hollows and crotches. Such plants relish the dampness found so close to the river, as well as the dense cover that surrounds them.

Nearby are the partial ruins of the fourteenth-century Rosslyn Castle, now restored in part as holiday accommodation. Wallace's Cave, thought to have sheltered Sir William Wallace and up to 70 soldiers after the Battle of Roslyn in 1303, is also close to the chestnuts. So is the fifteenth-century Rosslyn Chapel, founded in 1446; its amazing interior is now famous as a film location for *The Da Vinci Code*.

These two ancient trees can be found in Roslin Glen, just off the public path east of Rosslyn Chapel and north of the North Esk river. Work at the site by Midlothian Council's Land and Countryside Service has cleared ash and sycamore from around the chestnuts. This has helped to reduce competition and to provide more space and light.

LOCATION
Rosslyn Chapel,
Chapel Loan,
Roslin,
Midlothian,
EH25 9PU

OS GRID REF
NT278632

One of two ancient chestnuts close to the partial ruins of Rosslyn Castle, with huge burrs growing around its base.

45. SALTRAM SYCAMORE

Secrets of a sycamore

An outstanding old sycamore (or should that be sycamores?) greets you on arrival at the car park to the National Trust property of Saltram House and Gardens.

Much of the landscape at Saltram was designed in the eighteenth century around the Georgian house. Over time the landscape of the park has changed, with parts being divided up by Devonshire hedge banks. The grounds now form Saltram Countryside Park.

This very special multi-stemmed sycamore (*Acer pseudoplatanus*), now a central feature in the car park, raises many questions. These are hard to answer, as there seems to be no written evidence as to why this tree has developed its shape.

I have visited the Saltram Sycamore many times and reflected on its origins. I believe that it was once part of a hedgerow, long ago removed and perhaps layered. Admittedly, none of the old maps make this obvious, but the sycamore's exposed roots seem to indicate this.

The tree itself is made up of eight individual trunks, some of which divide again low down. Two of these have been lost, but a very impressive row of trunks remains. It is around 7.5 m (25 feet) in length with well-used gaps between each of the trunks, almost resembling purpose-built seating areas. Certainly this sycamore is a magnet for children, who run around and climb over between the trunks, although the long-term effect of this is sure to be detrimental to the tree. However, apart from its lost limbs the sycamore looks in remarkably good health.

Another possible reason for these trunks, all lined up but obviously joined as one, is that it might have been a fallen tree. Left to its own devices it layered itself, producing the multiple stems, and was then purposely left as a feature.

However this sycamore came to be, it is a wonderful natural creation.

LOCATION
Saltram House Carpark, Saltram, Plympton, Plymouth, Devon, PL7 1UH

OS GRID REF
SX522554

The remarkable, multi-stemmed Saltram Sycamore may be a remnant of an ancient hedgerow.

46. SAVERNAKE BEECH AVENUE

A living legacy

The beech-lined Grand Avenue is the longest avenue in Britain at just over 4 miles (6.5 km) in length. In its heyday, as old photographs show, each side was lined with many more, much larger beech trees than now exist. However, this route into the larger Savernake Forest still takes you back through history.

A magnificent avenue of beech trees of mixed ages, Grand Avenue runs right through the heart of Savernake Forest. Turning off the busy A4 on to this unclassified road, very little remains to show that this tree-lined avenue running north-west to south-east was once a far grander avenue. Two large, stone-faced gate piers of the old entrance are all that survive to indicate the history of this private road, which leads to the northern end of the estate. The avenue originally led all the way to Tottenham House (now just outside of the southern boundary of the forest), which still survives and is in private ownership.

Tottenham House was once owned by Charles Bruce, the Earl of Ailesbury, who may well have started the laying out of the avenue early in the eighteenth century. The design was then updated in plans laid out later in the century by Lancelot (Capability) Brown, with the addition midway along the avenue of the junction known as Eight Walks. Here it is said a gibbet, commonly used for executions in the eighteenth century, once stood as a macabre warning for highwaymen. At Eight Walks the road radiates out along tree-lined rides into one of the oldest forests in England; here more veteran trees grow than anywhere else in Europe.

The avenue may not be as spectacular as a century ago, when many more of the trees were more uniformly mature. Yet in its full autumnal splendour this beech-lined road, now consisting of many self-sown beeches, remains a great living legacy from a previous generation.

The avenue, still a private road, is closed one day every year to preserve this status.

LOCATION

Unclassified road off the A4: Grand Avenue, Savernake, Savernake Forest Mildenhall, Wiltshire.
Nearest postcode SN8 3HN.

OS GRID REF

SU210683

Running through Savernake Forest, beech trees of the Grand Avenue display magnificent golden colours in the autumn sunshine.

47. SAVERNAKE FOREST

Royal romance

This amazing forest in the county of Wiltshire was recorded as 'Safernoc' in a tenth-century charter of King Athelstan. A remnant of it now covers an area of approximately 1,800 hectares (4,500 acres). Once, however, it encompassed almost four times that area. Savernake Forest has a history dating back more than 1,000 years.

During the twelfth century it was established as a 'Royal Forest', a medieval forest ground for hunting royals. Henry VIII hunted stags and wild boar there in the sixteenth century. Sir John Seymour, warden of Savernake Forest, was the father of Henry's third wife Jane Seymour.

The forest is predominantly beech and oak. It is believed to have the highest concentration of veteran trees in Europe, with around as many as 2,500 veteran and ancient oaks and a similar number of beeches. Many of these trees, particularly the ancient oaks, are among the oldest in Britain. The older beech trees were planted during the seventeenth century, possibly in the reign of Charles I.

By the mid-twentieth century many of the old trees were suffering from being overcrowded and shaded out by the younger trees around them. The practice of 'haloing', or clearing space around older, vulnerable trees in a forest by removing others that are shading them out, was started during the 1970s, to help prolong the life of the forest's most treasured trees. Some of these are named, for instance the Cathedral Oak and the Big Belly Oak – two magnificent trees that stand on opposite sides of the forest. The Big Belly Oak, with its wide trunk, is almost 11 m (36 feet) in girth and held together with an old metal strap. This majestic tree's 'belly' bulges out beside (and almost over) the western edge of the forest along the A346 south of Cadley. Legend has it that dancing around this tree naked can summon the devil! This was also one of the trees included in the 'Fifty Great British Trees' that formed part of the Queen's Golden Jubilee celebrations in 2002. It is well worth admiring – but beware of the fast-moving traffic.

Another well-known named tree is sadly now all but gone; just one section remains upright. It is called Duke's Vaunt Oak. This ancient oak was illustrated and described in an edition of the *Gentleman's Magazine* in 1802 as a tree measuring 30 feet (9.1 m) on the outside, with a hollow measuring 20 feet (6 m) around the inside. The author of a letter published about the Duke's Vaunt Oak in the magazine writes

LOCATION
Savernake Forest
Marlborough,
Wiltshire.
Parking in Postern
Hill car park
SN8 4ND.

OS GRID REF
SU198679

The Big Belly Oak in Savernake Forest is bound with a metal belt to keep its bulging trunk intact.

I remember on a perambulation round the boundaries of that parish about 40 years ago, being one of twenty boys who were shut up together in the cavity; and also that a band of music consisting of a violin, houtboy (oboe), and bassoon, played several tunes in it.

Apparently a lockable door also hung on it.

There are many paths, tracks, and avenues, including 'The Grand Avenue', p.137) that bisect the forest. Some were established in the eighteenth century by the renowned landscaper Lancelot 'Capability' Brown. One, according to Peter Underwood, an eighteenth-century ghost hunter, is haunted by a headless woman riding a white horse; she is said to have been decapitated by a branch during a royal hunting party.

Like all ancient forests, Savernake is home to many rare flora and fauna that have developed along with them. It is also home to many rare and endangered species of lichen, most growing on the trunks of the ancient and veteran trees. Savernake is therefore recognised as a Site of Special Scientific Interest (SSSI). Some of these lichens are so rare that they are found in only a few sites in the south of England and Wales. Sweden is the only other country in which they occur.

To improve the area's biodiversity, White Park cattle have been introduced to part of the forest known as Red Vein Bottom. Here the cattle roam freely; they graze, remove woody seedlings and keep glades open, create 'wood pasture' areas where the regrowth is continually eaten and allow a habitat with light and space for some of the rarer flora to survive.

Savernake Forest, the largest privately owned ancient hunting forest in England, is owned by the Earl of Cardigan and his son, Viscount Savernake. The forest has continued to be managed in an unbroken line for 31 generations. In 1939 a lease was signed for 999 years to the Forestry Commission, assigning them the rights to the bulk of the timber and management of the forest.

Sitting on what is now the boundary of today's much reduced Savernake Forest is the magnificent Cathedral Oak, a huge pollarded tree.

The Saddle Oak at Savernake Forest somehow manages to support the vast weight of a huge limb almost at right angles to the main trunk.

48. SCONE DOUGLAS FIR

A living memorial

Although not perhaps the most exciting looking tree, the Scone Douglas Fir is significant for being one of the first cultivated in Britain, from seed collected in the nineteenth century from trees growing in its native habitat in the Pacific Northwest. Throughout its natural range, this versatile tree was well-used by Native American peoples for a variety of medicinal purposes, and for building and basketry. Commercially, Douglas Fir has become one of world's most important timber trees.

The tree's botanical name *Pseudotsuga menziesii* commemorates the Scottish botanist and naturalist Archibald Menzies, who documented the tree in 1791. But its common name recognises David Douglas (1799–1834). In April 1825, during his second collecting expedition for the Royal Botanic Institution of Glasgow, Douglas collected seed from trees growing along the Columbia River and sent them back to Scone.

David Douglas (1799–1834) was born close to Scone Palace, in the village of Scone in Perthshire. As a young man he worked as an apprentice for seven years in the Palace gardens. So it is fitting – and poignant – that a tree from the original batch of seed he sent from his North American expedition still thrives there today. In 1834, during his third expedition, Douglas died in Hawaii at the age of only 35. Accounts of the cause of death give various reasons, but one claims that he fell into a wild bull pit, already occupied by a wild bull. Such an early death was a tragic loss to botany and natural science.

The Scone Douglas Fir exhibits the dark brown bark, thick and deeply fissured, that typically clothes the large tapering trunk of this species. Another distinctive feature is the large lower limb that heads towards the ground before curving back upwards, creating a long, sweeping, trunk-like arm. This individual fir provides a living memorial to Scotland's greatest plant collector, who introduced over 200 species of plants to British horticulture. A quote from the *Gardener's Chronicle* in 1926 reads 'to no single individual is modern horticulture more indebted than to David Douglas.'

LOCATION
Scone Palace,
Perth,
PH2 6BD

OS GRID REF
NO113265

One of the Douglas firs grown from seed collected by David Douglas in the nineteenth century. The tree still thrives in the gardens at Scone Palace.

49. SHERWOOD FOREST

Outlaws hideout

The mention of the name Sherwood Forest will for many conjure up stories of Robin Hood, the legendary outlaw who, backed up by his band of merry men, stole from the rich to give to the poor. As legend would have it, they used the forest for their hideout and concealed poached game in some of the old hollow trees. One such tree was named the Larder Oak, although there is no real evidence that it was used in this way.

Stories of a Robin Hood character, famed for his skills as an archer, go back as far as the thirteenth century and have long put Sherwood Forest on the map. For centuries forests were a source of refuge, harbouring criminals and those on the run. The main route from London to York, the Great North Way, ran right through Sherwood Forest, making it an ideal place for outlaws and highwaymen to surprise, attack and rob passing travellers.

The earliest name for the forest was *Sciryuda*, meaning 'the woodland belonging to the shire'; the name Sherwood was first recorded in the tenth century. Like many of England's large areas of woodland, it became a royal hunting forest after the Norman Conquest in 1066. Once woodland was deemed a royal forest it came under royal law, which overrode any earlier common law. The vert and venison (timber and game) now belonged to the ruling monarchs, and hunting was only allowed with royal permission.

At this time Sherwood Forest covered a large part of Nottinghamshire, around 40,470 hectares (100,000 acres). Much of this was heath and open grassland, which in medieval times would have been managed as wood pasture; here cattle and sheep were grazed, and pigs fattened up on the autumn harvest of acorns. Trees were coppiced, pollarded and used for building material, fuel, and charcoal production, while the bark of the oak was used for tanning leather.

In his book published in 1875, *Worksop, The Dukery and Sherwood Forest*, Robert White notes several accounts of the trees being felled, mainly for timber:

> *In a survey again made in 1790, it appears that there were then in Birkland and Bilhagh together only 10,117 trees, which at that time were valued at £17,147 15s. 4d., so that in 104 years (from 1686–1790) 27,199 trees had gone.*

LOCATION
Sherwood Forest,
Edwinstowe,
Nottinghamshire,
NG21 9RN

Parking in Sherwood main carpark

OS GRID REF
SK626670

The most famous oak in Sherwood Forest, the Major Oak, is reputed to have links with the legendary Robin Hood.

The eighteenth century also saw the landscape transformed in other ways:

> *Several Inclosure Acts were passed between the years 1789 and 1796, for Arnold Forest, Basford, Sutton-in-Ashfield, Kirkby-in-Ashfield, Lenton, and Radford, and thus about 8,248 acres were brought into cultivation. By these means it has come to pass, that in this district, with the exception of Birkland and Bilhagh, and a few ancient trees scattered here and there, and besides what are included in the parks of the nobility and gentry, scarcely a vestige of the Ancient Forest of Sherwood now remains except in name.*

Today what remains of this once huge forest is a smaller area of around 182 hectares (450 acres). It is known as Birklands (a Viking name meaning 'Birch Land'), which became 'the King's hay of Birkland' in the seventeenth century.

Sherwood Forest is almost entirely made up of oak and birch, with the former dominating the woodland – many are centuries old. Around 900 ancient and veteran oaks are recorded throughout the forest, one of the richest in the country for surviving old oaks. Today it is both a National Nature Reserve (NNR) and a Site of Special Scientific Interest (SSSI).

The tree for which Sherwood Forest is best known today is the Major Oak, thought to be at least 1,000 years old. It has a girth of over 10 m (33 feet), and its huge canopy spreads out to an impressive 28 m (92 feet). Through its long history the Major Oak has been known by various names. It was called the Cockpen Tree in the mid-eighteenth century, a time when cock fighting was a sport carried out in the forest; presumably the open space that must always have surrounded this tree (hence its wide spreading canopy) made it an ideal place for gatherings. It was also known as the Queen's Oak, only becoming known as the Major Oak when featured in a book about the ancient oaks of Sherwood Forest, *Remarkable Oaks in the Park at Welbeck*. The author, a Major Hayman Rooke, lived locally and regularly walked in the forest; he too was known as the Major.

In the late eighteenth century many other ancient named oaks grew in the forest; sadly none of these remain today. However, many are captured in old drawings and photographs such as the Greendale Oak, with an arch cut through its hollow centre large enough to drive a horse and carriage through. Also notable were a large, coppiced stool with seven large trunks, known as the Seven Sisters, and Robin Hood's Larder Oak, mentioned above.

Another old postcard image of an oak with links to Robin Hood. The Larder Oak was supposedly used as a hideout for stolen goods or poached game.

An old image of the Major Oak from the early twentieth century helps to establish the age of this ancient oak.

50. ST JOHN THE BAPTIST CHURCH – YEW AVENUE

Corpse avenue

This amazing avenue may only consist of eight yews, but is probably the oldest avenue of yew in Britain. It is certainly the oldest in any churchyard. Suggestions regarding the age of these old yews, and when they were planted, emerge from various sources. An information board at the entrance of the thirteenth-century church states that they were planted at the end of the fifteenth century, when much of the church was rebuilt. There is also a suggestion that the trees were planted by Henry Fitzalan, the 12th Earl of Arundel, in the mid-sixteenth century, but again exact dates vary. Taking a date in between those proposed would give the yews an age of around 500 years.

From the north gate the short avenue runs south to the church doors, flanked by four opposite pairs of trees of fairly similar sizes. The yew closest to the church porch has the largest girth. Only one of the eight trees, the first yew on the left just inside the churchyard gate, is female. Its girth is marginally the smallest. All have beautiful, deeply fluted red-brown trunks.

Avenues of yews were often planted from the lychgate (also known as corpse gate) to the church door. The lychgate was a sheltered area in which coffins were rested on their way into church while awaiting the clergyman's arrival. The evergreen yew trees, as well as being associated with the dead, gave year-round shelter along the funeral route, lining the path to the church entrance.

LOCATION
Westbourne Road,
Westbourne,
Emsworth,
West Sussex,
PO10 8UL

OS GRID REF
SU755073

This avenue of eight yews with fluted trunks leads through the churchyard to the chapel doors of St John the Baptist Church. Though short, it is probably the oldest avenue of yews in Britain.

51. STIPERSTONES HOLLIES

Life in the old hollies yet

The Stiperstones Hollies are a remnant grove of ancient hollies (*Ilex aquifolium*). They are scattered over the northern end of the Stiperstones ridge, an area that would have once had many more of these amazingly characterful old trees. Holly woods were also known as 'Hollins'.

Some of the trees that survive in the grove are thought to be over 400 years old and among the oldest in Europe. The open and windy ridge is home to many of the survivors, and it is this exposed site that has given the hollies their individually gnarled and wizened, windswept shapes. Old they may be, but the female hollies still produce an abundance of their bright red berries in late autumn, attracting flocks of birds who stock up on this seasonal feast.

Many rowans also grow among the hollies. Birds feeding on their berries sometimes inadvertently sow the seeds in the decaying parts of the old hollies as they perch or eat in their branches, leading to a phenomenon known as 'cuckoo trees'. It is quite common to see these rowans growing out of their holly hosts.

The area below the ridge was once mined for its rich seams of lead, from shafts over 300 m (more than 1,000 feet) deep in the Snailbeach lead mines. The miners lived with their families up on the ridge among the hollies, and it is partly the protection they gave the trees, cut back above the browsing-line to produce nutritious winter fodder for livestock, that has prolonged the hollies' lives. Their dwellings have long gone, but some grazing continues with sheep and cattle. Today many of these ancient hollies are fenced for their protection, to allow the trees to thrive for years to come.

The Hollies is a nature reserve covering almost 36 hectares (90 acres). It is managed by the Shropshire Wildlife Trust.

LOCATION
Lord's Hill,
Snailbeach
Minsterley,
Shropshire,
SY5 0NS

OS GRID REF
SJ375022

Protected from grazing, this and other hollies now sparsely spread along this ridge were once part of a much larger grove at Stipersones. Some of the trees in the grove are thought to be over 400 years old.

52. TOLPUDDLE SYCAMORE

Meeting place for the martyrs

The Tolpuddle Sycamore may not be the largest sycamore in Britain but at 5.5 m (18 feet) its girth still puts it in the exceptional class. And together with an age of more than 300 years, this veteran is certainly the most famous sycamore and one of the most famous trees in the UK. For it was under this tree in 1834 that six poverty-stricken farm workers earning just nine shillings a week (equivalent to around £10 a week today) decided to group together and form a 'friendly society' to fight for better pay and conditions, and to stop their employers making further cuts to their pay.

These six farm workers were barred from using indoor spaces for their meetings to discuss their low pay and long hours, which were considered 'unlawful' by the authorities of the time. They are thus believed to have used the old sycamore at the edge of the village (which would have already been a substantial tree, around 150 years old), as a meeting point. The six men (George Loveless, James Loveless, James Brine, James Hammett, John Standfield and Thomas Standfield) became known as the Tolpuddle Martyrs.

The group were arrested and put on trial for trying to form a union. For their 'crime' they were sentenced to seven years' transportation to Australia.

A rally attended by tens of thousands of protesters gathered and marched to Parliament. Here they handed in a petition to have the men pardoned and returned. However, it was not until March 1836 that the Tolpuddle men were finally pardoned and allowed to head back to England as free men. George Loveless was the first to arrive home in June 1837; the last was James Hammett in August 1839. Their story is celebrated every year in the village of Tolpuddle with a festival in July.

Growing on what must be one of the smallest village greens in the UK, the Tolpuddle Sycamore stands as a memorial to the early struggles of workers to found unions and improve their conditions; the evocative site is considered the birthplace of the Trade Union movement. Regular pollarding should now help this iconic tree, known all around the world, to survive for another couple of centuries at least. However, a young tree grown from a seed of the original tree has been planted on the other side of the green, in case its parent should fail.

The sycamore is now cared for by the National Trust, as the green on which it grows was given to the Trust in 1934 by Sir Ernest Debenham.

LOCATION
Tolpuddle,
Dorchester,
Dorset,
DT2 7EX

OS GRID REF
SY791944

The Tolpuddle Sycamore, believed to be the meeting place where the Tolpuddle labourers met to demand better wages and working conditions, is historically the most celebrated sycamore in Britain.

53. TORTWORTH CHESTNUT

Family gathering

This ancient sweet chestnut (*Castanea sativa*) makes up a mini woodland all on its own. Its imposing gnarled structure stands in a fenced-off compound, within a small field to the right of St Leonard's Church.

Calculating the age of trees such as this, with its huge trunk made up of a gnarly mixture of living, decaying and dead material, is an almost impossible task; guesstimates range from 400 to 800 years old. The Tortworth Chestnut is certainly a remarkable tree, believed to be one of the oldest sweet chestnuts in Britain.

An inscription on a brass plaque, mounted on the gate close to the tree, observes:

> *THIS TREE IS SUPPOSED TO BE*
> *Six Hundred Years Old*
> *1st Jan^{y}, 1800*
> *May Man Still Guard thy Venerable form,*
> *From the Rude Blasts and Tempestuous Storm,*
> *Still mayest thou Flourish though Succeeding time,*
> *And Last, Long Last, the Wonder of the Clime.*

We know that the tree was measured on various occasions from the early to mid-eighteenth century. A circumference of over 50 feet (15.2 m) was recorded. In 1762 the chestnut was measured at 52 feet (almost 16 m) by amateur dendrologist Peter Collinson, who declared that the chestnut was 'in all probability the oldest, if not the largest tree in England'. Collinson went on to suggest that it may have grown from a nut in the reign of the Saxon king Egbert, around CE 800.

It is also claimed that the chestnut may have been a boundary marker in the twelfth century for the Tortworth Estate, in which it still sits today. This tree has certainly been well documented through the centuries, though how much of this is factual remains open to debate.

Today, the Tortworth Chestnut is a slimmed-down version of what it may have been in its prime (although drawings and engravings from the eighteenth century portray it even then as quite an ancient tree in decay). The most recent measurement around what remains of the original trunk is still approximately an impressive 12 m (39 feet). And when you include all of the sibling clones with which the tree has surrounded itself inside its protective compound, it clearly has an impressive legacy.

LOCATION
Near Saint Leonard's Church, Tortworth, Wotton-under-Edge, Gloucestershire, GL12 8HF

OS GRID REF
ST704932

A single tree (main trunk on the right), the layered limbs of the Tortworth Chestnut have become individual large trees, creating a small, united family copse.

54. WAREHORNE ASH

A venerable ash

In the churchyard of Warehorne's beautiful Grade I listed medieval stone church (apart from the tower, which was struck by lightning in 1777 and rebuilt in brick) stands an ancient ash tree (*Fraxinus excelsior*).

A church on the site of the existing church is mentioned in the Domesday Book of 1086. So there must have been a much earlier building than the church of St Matthew that stands on the site now and dates from the thirteenth century.

The ancient ash in the churchyard of this small village cannot claim to be as old as the church. However, it is thought to be around 500 years old, which for an ash tree is very old indeed. This remarkable tree will have seen many generations of worshippers, wedding guests and other congregations visiting the church. Today it shares the old churchyard with a small flock of sheep, which keep the grass in check and the old tree company.

The Warehorne Ash has been pollarded over the years and is now completely hollow. It is basically in two main sections, so you can see right through the tree. The two sections are held together with a cable bracing system, fitted in 2018 in an attempt to extend the old tree's life and stop it literally falling apart. Sadly, although the tree has continued to produce new growth each year, it has more recently been showing some signs of ash dieback.

LOCATION
Church Road,
Warehorne,
Ashford, Kent,
TN26 2LP

OS GRID REF
TQ989325

Hollow, fragmented and held together with cables, the valiant Warehorne Ash looks every bit five centuries old.

55. WESTONBIRT LIME

Circle of life

This ancient small-leaved lime (*Tilia cordata*) is thought to be between 1,000 and 2,000 years old and is certainly among the oldest trees in Britain. Trees of this age mostly tend to be yews and, to a lesser extent, oaks. This lime, now made up of a circle of individual stumps known as stools, has been coppiced, or continually cut back to ground level, for millennia; this has brought about its longevity. DNA work carried out on the lime in the 1990s proved that all the stools originate from one original tree.

The ancient circle of limes grows in an area of the Westonbirt Arboretum known as Silk Wood. Dating back to the thirteenth century, it was once wild woodland, becoming the last area of the estate to be developed into part of the arboretum in the late nineteenth century. Close by is the arboretum's main collection of maples, providing the riot of the autumn colour for which Westonbirt is famed.

Around 20 individual stools make up the growth above the ground, which measures approximately 25 m (82 feet) across; it is from this figure that the lime's age can be calculated. These individual stools will all be linked below ground. They continue to share resources as though they were still a single tree, as they have done ever since they were a seedling.

The Westonbirt Lime is coppiced around every 20 years, most recently in 2012. The following year, in March 2013, cut stems from that coppicing were used by the artist Richard Harris. He was commissioned by the Westonbirt Arboretum, on behalf of Forestry England, to produce a celebratory sculpture of its ancient lime.

Several tons of the cut stems were arranged vertically within a steel frame to represent the stem of the lime if it had been a single trunk, complete with a hollow centre. The 'reconstructed tree' had a circumference of 15 m (49 feet) and was 11 m (36 feet) high, bringing to life the lime's possible scale and size if it had survived as a single tree.

Westonbirt Arboretum covers 240 hectares (600 acres) and contains a tree collection from around the world. It has an admission charge (check online) and includes a fascinating range of 'champion trees', the largest, or tallest, of their kind in Britain. The ancient lime is around a 30-minute stroll from the main entrance.

LOCATION
Westonbirt
National Arboretum,
Westonbirt,
Tetbury,
Gloucestershire,
GL8 8QS

OS GRID REF
ST854896

A circle of coppiced stools have continued the life of the Westonbirt Lime, continuing to rejuvenate it for more than 1,000 years.

56. WINDSOR GREAT PARK

Royal oaks

LOCATION
Cranborne
Gate car park,
SL4 2BY

OS GRID REF
SU978705

The old Forest of Windsor once covered a huge area, stretching out of the region of Windsor and into the surrounding counties of Berkshire, Buckinghamshire and Surrey. It was a hunting forest as far back as the eleventh century. In Addlestone a tree known as the Crouch Oak is thought to be a tree of the old Windsor Forest's Surrey boundary; it has now become a street tree on the outskirts of the town.

Today Windsor Great Park consists of only a small portion of the original forest. It covers 2,020 hectares (5,000 acres) and is bisected by the A322, with the largest area to the south of this road. This part of the Great Park has seen the most development, through the landscaping of areas such as The Savill Garden, Valley Gardens and Virginia Water, originally dammed and flooded in 1753. At the time Virginia Water was the largest artificial lake in Britain.

Cranbourne Chase (some of which is private) lies to the north of the A332, just outside the boundary of Windsor Great Park, but still part of the Crown Estate. Once part of the larger Forest of Windsor, the area is home to some of the many ancient and veteran trees spread across the Great Park.

William the Conqueror was the first to construct a home on land within the forest, on high ground above the River Thames. Though not as elaborate as the castle that stands today, parts of Windsor Castle still date from that period and so are nearly 1,000 years old. The building has been a royal palace and home to British monarchs throughout the centuries; it is currently the official residence of Her Majesty The Queen. Amazingly, some of the ancient oaks in the surrounding grounds date back to the origins of the castle; some are even older.

The castle was built on the edge of a Saxon hunting ground; it continued to be used for the 'Royal Chase' up until the eighteenth century. Today parts of the park are still managed as a deer park, but not for hunting purposes.

There are well over 100 ancient and veteran oaks within Windsor Great Park – many of them remnants of the wild wood. Most are well known and many of the significant trees have names. The Signing Oak, for example, is thought to be around 900 years old. It has a girth of almost 10 m (over 32 feet), the second largest of any of the ancient oaks in the Park. Growing in an area of wood pasture, the Signing Oak enjoys the company of other ancient oaks nearby.

The oldest living tree in the park is an English, or pedunculate, oak. It

William the Conqueror's Oak at Windsor Great Park. Now completely hollow, the tree is propped and circled by a low fence for its protection.

is named King Offa's Oak, after the eighth-century Mercian king, and is said to date back to his reign in CE 710. Possessing a girth of over 11 m (36 feet), it also has the largest girth of the ancient oaks in the Park. Not surprisingly, the centuries have taken their toll on this magnificent old tree. Two individual sections are propped and partially collapsed, and a third section lies on the ground. However, King Offa's Oak still has lots of healthy parts that produce new growth every year.

Close by is another named oak, William the Conqueror's Oak. This tree is completely hollow, with a large section of deadwood in the trunk. The Conqueror's Oak, like Offa's Oak, is supported by metal props and ringed by a low fence, protecting it from soil compaction around its roots.

In the mid-nineteenth century William Bambridge, photographer to Queen Victoria, captured images of a series of oaks named after queens. Queen Charlotte's Oak was a tree with a short trunk and wide girth; it had many tall limbs which may have been a lapsed pollard. Queen Anne's Oak was a large maiden (an oak that has grown naturally, without being pollarded: they have a more rounded, 'natural' canopy that is evenly branched). The most famous of these queen oaks, Queen Victoria's Oak, was another tall maiden, its trunk flaring towards its base. The form of this tree inspired the building of the Plymouth lighthouse. Engineer John Smeaton was impressed by the tree's shape, and the way in which it flexed in strong winds, and so he designed the lighthouse structure based on this tree. This was the third lighthouse to be built on the Eddystone Rocks, where it was also known as Smeaton's Tower. It was used for over 100 years before deterioration of the rock beneath made it unstable. It was replaced in 1878, with the old Smeaton's Tower being rebuilt on dry land at Plymouth Hoe.

As with many ancient woods, forests and trees, there is plenty of myth, mystery, and folklore in Windsor Great Park. It even has its own ghost: Herne the Hunter. The mythical hunter rides through the park on horseback on stormy nights, wearing a set of stag's horns on his head. In Shakespeare's play *The Merry Wives of Windsor*, Herne the Hunter is impersonated by the jovial Falstaff. An oak was named after him, Herne's Oak, which was already ancient oak in Shakespeare's time. Paintings and drawings in the eighteenth century show the tree in decline, with just a few branches still alive.

One of the many beautiful old oaks in Windsor Great Park, this is King Offa's Oak.

57. WISTMAN'S WOOD

A lonely, bewitching wood

When I first walked into this magical and beautiful woodland many years ago, I was stopped in my tracks by the sight of the dwarf oaks. Their gnarled and entwined branches twisted in all directions, and below them an understorey of moss and lichen covered granite boulders. The changing dappled light creates dancing shadows and patterns across the boulders, known locally as clitter, adding to the wood's spectral qualities. The branches are also clothed in a soft, verdant covering of epiphytic mosses and hanging lichens. Wistman's Wood is truly an amazing, mystical and magical place, not surprisingly associated with much folklore and many stories. One of the legends tells of the wood being home to large black hounds, 'Wisht Hounds', with blood-red eyes and huge yellow fangs. These terrifying creatures roam the moors on dark and misty nights looking for unwary travellers; they are said to be led by the Devil, or by the Old Crockern, an ancient spirit that lives on the moor on Crockern Tor.

Wistman's Wood is one of three isolated ancient upland woodlands on Dartmoor. It sits on the eastern slopes of the West Dart river in the Dartmoor National Park. The wood is managed by the Nature Conservancy Council and owned by the Duchy of Cornwall. It is one on the highest oak woodlands in Britain, about 405 m (1,330 feet) above sea level, and has been a Site of Special Scientific Interest (SSSI) since 1964. It is a remnant of around 3.5 hectares (almost 9 acres) of what was once a much larger wood.

The trees are mainly *Quercus robur*, the English (or pedunculate) oak, and have a dwarf habit, reaching around 4 or 5 m, occasionally 6 m (10 or 12, occasionally 15 feet) in height – no doubt a consequence of poor soils and the strong winds that whistle over the moor. Their procumbent trunks and branches sometimes lie between the boulders on the ground of this miniature woodland. However, there are also rowans or mountain ash (*Sorbus aucuparia*), as well as hollies (*Ilex aquifolium*), and more than 100 species of lichen have been recorded here. Rocky boulders and uneven ground have protected the trees from grazing animals, particularly Dartmoor ponies.

Tristram Risdon, an English antiquarian and topographer of the seventeenth century, wrote in his *Survey of the County of Devon*:

LOCATION
Two Bridges,
Princetown,
Dartmoor,
Devon,
PL19 9NA

OS GRID REF
SX612774

The whole of Wistman's Wood sits in isolation on the eastern slopes of Dartmoor National Park.

The stunted oaks and stone boulders of Wistman's Wood, cloaked in mosses and lichens. A truly magical place, it is linked to many myths and folk tales.

In this forest are three remarkable things: The third is some acres of wood and trees, that area fathom about, and yet no taller than a man may touch to top with his hand, which is called Wistman's Wood.

On the outskirts of the wood is a large triangular lichen-covered stone, known as the Buller Stone. It bears an inscription commemorating the removal of an oak on 16 September 1866 by Wentworth Buller. The tree measured just 23 cm (9 inches) in diameter and was estimated to be about 168 years old, showing how slowly the trees grow in this harsh environment. Wistman's Wood is a short walk (around 1.6 km or 1 mile) from a car park opposite the Two Bridges Hotel on the B3357 in the heart of Dartmoor.

How heavily
That old wood sleeps in the sunshine; — not a leaf
Is twinkling, not a wing is seen to move
Within it; — but, below a mountain-stream,
Conflicting with the rocks, is ever heard,
Cheering the drowsy noon.
of this grove,
This pigmy grove, not one has climb'd the air
So emulously that its loftiest branch
May brush the traveller's brow. The twisted roots
Have clasp'd, in search of nourishment, the rocks,
And straggled wide, and pierced the stony soil: —
In vain, denied maternal succour, here
A dwarfish race has risen. Round the boughs,
Hoary and feeble, and around the trunks,
With grasp destructive, feeding on the life
That lingers yet, the ivy winds, and moss
Of growth enormous. E'en the dull vile weed
Has fixed itself upon the very crown
Of many an antient oak; and thus refused
By Nature kindly aid – dishonoured – old –
Dreary in aspect – silently decays
The lonely Wood of Wistman.

Nicholas Toms Carrington (Poet 1777–1830)

58. WOODSTOCK GARDENS AND ARBORETUM – NOBLE FIR AVENUE

Noble avenue

Originally this avenue was known as the Silver Fir Walk, as the trees were thought to be *Abies alba*. However, they were later correctly identified as noble firs (*Abies procera*), so the avenue has been renamed Noble Fir Walk.

Planted in 1878, the firs are set in 44 opposite pairs on each side. They are spaced around 7 m (23 feet) apart along the length of the avenue, which is roughly a quarter of a mile long (a little under half a kilometre). The magnificent survivors of the original planting are an awesome sight, towering to almost 35 m (115 feet); the largest trees have girths of over 4 m (13 feet). Walking along between them, you can be forgiven for not associating these noble firs with one of the most popular species harvested each year: Christmas trees. Their true beauty and scale are only achieved with maturity.

The avenue was planted under the direction of Lady Louisa Tighe and William Frederick Fownes Tighe, Lord Lieutenant of Kilkenny and privy councillor. William died in the year that the firs were planted, so unfortunately never got to see the avenue take shape. Yet he and Lady Tighe would surely have been delighted to see their legacy today – the stunning, tree-lined walk they surely envisaged nearly 150 years ago.

The arboretum at Woodstock has another avenue, also planted in the nineteenth century, but earlier than the firs. It was originally planted with monkey puzzles (*Araucaria araucana*), in 1845. The early plantings struggled in the wet soil conditions, so many were re-planted in the 1860s, in improved soil conditions. Like the later-planted Noble Fir Avenue, the Monkey Puzzle Walk is an avenue lined with beautiful giants, the tallest reaching over 28 m (92 feet). The seeds these bear are being used to grow new trees; these in turn will replace some that are lost and restore this amazingly captivating walk.

LOCATION
Woodstock Gardens and Arboretum, Inistioge, Co. Kilkenny

OS GRID REF
SL565984

These magnificent towering conifers are what make Noble Fir Walk such an amazing feature avenue.

59. WOOLLAND YEW

A magnificent old lady

The wonderous Woolland Yew sits in the corner of the small churchyard of St Mary's Church. A huge female tree, it takes up much of the south-west side of the churchyard, close to the pretty little church's porch. The yew is almost separated into five large trunks from low down; from the rear it appears more like a single, huge, deeply fluted trunk.

From the front (the side nearest the small entrance gate) there is a large open area. Once presumably a hollow, it has over time become completely open and large enough to enter. An old metal bar high in the canopy holds some of the central limbs together, attempting to help strengthen them and prevent them pulling apart.

John Hutchins's *History and Antiquities of Dorset*, published in 1861, records that this ancient yew had a measured circumference of 23 feet (7 m). Its most recent measurement, taken in 2012, is 31 feet (nearly 9.5 m).

A certificate in the church porch certifies the yew as being 2,000 years old. This is probably an exaggeration, but the Woolland Yew is indeed a remarkably ancient tree.

Interestingly, four documented churches in Woolland have been built on this site since the fourteenth century, but it is believed to have originally been a Saxon site. The current church, built in 1856, includes recycled material from the earlier churches. Carvings depict the leaves of every tree in the village of Woolland (although some more exotic plants probably grow in the village now that would not have been there in the nineteenth century). The church also features a stone replica of a robin's nest to the right of the east window. It represents one that was apparently left undisturbed while the church was being built.

This incredible yew has been here through all of these changes.

LOCATION
St Mary's Church,
Hill Farm Lane,
Woolland,
Dorset,
DT11 0EP

OS GRID REF
ST776069

Churchyard yews often have a single huge hollow trunk. This one is unusually multi-stemmed, but still hollow and possibly fragmented.

60. WYNDHAM'S OAK

A peaceful old age

In a small, tranquil meadow behind the parish church of St Nicholas, parts of which date back to the thirteenth century, grows an ancient pedunculate, or English, oak (*Quercus robur*). The oldest oak in Dorset, this tree may even predate the oldest parts of the church.

Most commonly known as Wyndham's Oak, it is also called Silton Oak and Judge's Oak. Now standing in the meadow with few other trees, this ancient oak is thought to have been a boundary tree on the edge of the forest of Gillingham. Once a royal hunting forest, the area was deforested for its timber between the fifteenth and seventeenth centuries. It might be that Wyndham's Oak survived because even then its large trunk was probably hollow, making it useless as a timber tree. We will never know – but fortunately, survive it did.

The oak gets its name from Sir Hugh Wyndham, who held the office of Judge of the Court of Common Pleas; he also served as a Fire Judge, dealing with compensation claims following The Great Fire of London in 1666. Sir Hugh bought the Silton Estate, in which the oak stands, in 1641 and lived close by in Silton Manor. Tales of the time describe how the judge would enter within the tree and smoke his pipe, or seat himself under its canopy to contemplate his cases. Judge Wyndham died in 1684 and is commemorated in the church, which contains a life-size statue of him.

Darker tales include the tree's association with the Monmouth Rebellion, a seventeenth-century uprising that sought to overthrow James II in favour of the Duke of Monmouth (one of Charles II's illegitimate sons). Wyndham's Oak may have served as a 'hanging tree', a gallows or gibbet to hang convicted rebels.

Having undoubtedly lived through many troubled centuries, the oak now enjoys its twilight years peacefully in its meadow. Its hollow trunk has an impressive girth of nearly 11 m (36 feet), held together with a continuous chain. The tree sits on wooden blocks around the upper part of the bole, cushioning it from the chain; its main limb is supported by a large wooden pole. Despite this apparent frailty, the oak's small canopy produces a healthy head of leaves each year.

A sign points the way to Wyndham's Oak, which is found after passing though a gate and stile at the side of the church. Limited parking is available.

LOCATION
Silton,
Gillingham,
Dorset,
SP8 5PR

OS GRID REF
ST782293

OPPOSITE
As with most ancient oaks, Wyndham's Oak is a completely hollow shell, with a relatively healthy canopy.

OVERLEAF
Oak woodland, reminiscent of the royal hunting forests of centuries ago.

INDEX

Images are indicated by page numbers in **bold**

PREVIOUS SPREAD
Autumnal colours of a beech-lined ancient trackway in West Sussex.

OVERLEAF
One of the many characterful ancient oaks in Windsor Great Park.

Hawthorn surviving in the toughest of conditions, amongst exposed limestone pavement.